MOST
DELICIOUS
POISON

MOST DELICIOUS POISON

THE STORY OF NATURE'S TOXINS— FROM SPICES TO VICES

NOAH WHITEMAN

Little, Brown Spark
New York Boston London

Little, Brown Spark
Hachette Book Group
1290 Avenue of the Americas, New York, NY 10104
littlebrownspark.com

First Edition: October 2023

Little, Brown Spark is an imprint of Little, Brown and Company, a division of Hachette Book Group, Inc. The Little, Brown Spark name and logo are trademarks of Hachette Book Group, Inc.

The publisher is not responsible for websites (or their content) that are not owned by the publisher.

The Hachette Speakers Bureau provides a wide range of authors for speaking events. To find out more, go to hachettespeakersbureau.com or email HachetteSpeakers@hbgusa.com.

Little, Brown and Company books may be purchased in bulk for business, educational, or promotional use. For information, please contact your local bookseller or the Hachette Book Group Special Markets Department at special.markets@hbgusa.com.

Illustrations by Julie Johnson

ISBN 9780316386579
LCCN 2023935281

Printing 1, 2023

LSC-C

Printed in the United States of America

To Shane

Contents

MOST DELICIOUS POISON

Introduction

A deadly secret lurks within our refrigerators, pantries, medicine cabinets, and gardens. Scratch beneath the surface of a coffee bean, a red pepper flake, a poppy capsule, a *Penicillium* mold, a foxglove leaf, a magic mushroom, a marijuana bud, a nutmeg seed, or a brewer's yeast cell, and we find a bevy of poisons.

The chemicals in these products of nature are not a sideshow—they are the main event, and we've unwittingly stolen them from a war raging all around us. We use these toxic chemicals to greet our days (caffeine), titillate our tongues (capsaicin), recover from our surgeries (morphine), cure our infections (penicillin), mend our hearts (digoxin), bend our minds (psilocybin), calm our nerves (cannabinol), spice up our food and drink (myristicin), and enhance our social lives (ethanol).

You might be thinking that to call these chemicals poisons or toxins is an exaggeration. After all, at the doses we typically use—a sprinkle, a tablet, a glass—these substances can improve our health and well-being. But at higher doses, as anybody who has ever had a hangover can confirm, these chemicals, whether directly or indirectly, can harm us, too. As sixteenth-century Swiss physician Paracelsus noted, "the dose makes the poison."

Paracelsus's maxim is perhaps too general to be useful—and maybe that was his point. It is hard to define a poison or a toxin. That ambiguity is part of the story too (I use the terms *poison* and *toxin* interchangeably in this book because their meanings largely overlap). At the wrong dose, even oxygen can be toxic. But there's a reason we don't call oxygen a toxin: plants and other organisms with chloroplasts don't produce oxygen to harm other organisms. The gas is simply a by-product of photosynthesis—the ability to turn carbon dioxide and water into sugar.

The chemicals that I call toxins or poisons, on the other hand, often function as weapons in what Charles Darwin called "the war of nature," that is, the struggle all organisms endure to survive and reproduce. Some of these struggles are mediated through interactions between organisms, for example, between predator and prey or plant and pollinator. Darwin mused about how these interactions themselves arose through coevolution: "It is interesting to contemplate a tangled bank, clothed with many plants of many kinds, with birds singing on the bushes, with various insects flitting about, and with worms crawling through the damp earth, and to reflect that these elaborately constructed forms, so different from each other, and dependent on each other in so complex a manner, have all been produced by laws acting around us."

One of the laws Darwin formulated was *evolution by natural selection*. Natural selection acts on heritable differences between individuals to improve their odds of survival or reproductive output over time. This type of evolution produces new adaptations. Although he focused on traits that he could see with his own eyes, like the varied beaks of the Galápagos finches that now bear his name, we now know that evolution, primarily through coevolution between species, has also generated a profusion of toxic chemicals hidden inside many different organisms. These organisms use the chemicals to gain the upper hand, through offense and defense, in the Darwinian struggle for existence that has played out since life itself began.

This book explores the fascinating and sometimes surprising ways that toxins from nature arose, have been used by us humans and other animals, and have consequently changed the world. We will follow several interrelated threads, or approaches, as we examine how these chemicals have influenced evolution and how they have penetrated each human life, for better and for worse.

One thread concerns the origin stories of the toxins that occur naturally in many organisms. These chemicals, counterintuitively, can help explain why the planet is so filled with life. This is because the war of nature that largely hinges around these chemicals is a dynamo that generates new traits and new species through cycles of defense and counterdefense between ecologically interacting species.

We will trace important similarities between how animals and humans co-opt the same toxins from other organisms and use them as tools of their own to improve their odds of survival and reproduction. This similar behavior reveals that humans, although special in so many ways, are just one of many species that use the chemicals in nature's pharmacopoeia — and that all creatures depend on this trove of toxins one way or another.

Throughout the book, we will learn how numerous plants and fungi and even some small animals produce copious amounts of toxins that mimic human hormones and neurotransmitters or block their function. On the flip side, you may be surprised to learn that our bodies produce small quantities of some of the strange-sounding chemicals that plants use as defensive shields — such chemicals as aspirin-like molecules and morphine. I will explain the physiology of this process in the human body and show how it can help us understand the human susceptibility to addiction. At the same time, the most promising new treatments for some of these substance use disorders come from the natural pharmacopoeia in the form of psychedelics. A closer look reveals that human use of these psychedelics is not new at all and can be traced to the ancient and ongoing practices of various Indigenous and local peoples across the planet.

Another thread follows medieval Europe's obsession with nature's toxins in the form of Asian spices, a hunger that motivated the Age of Exploration. The desire for new sources of spices, and for control of the flow of spices, triggered a geopolitical cataclysm that shaped the past five hundred years of human history and continues to do so today. One consequence, at least in part, is the global biodiversity and climate crisis we face.

As excited as I was to weave all these threads together and tell the story of nature's toxins, this is not what motivated me to write the book. Instead, the sudden death of my father under tragic circumstances stemming from a substance use disorder in late 2017 is what pushed me to embark on this project.

His long struggle with nature's toxins came to a head just as my collaborators and I uncovered how the monarch butterfly caterpillar resists the deadly toxins made by the milkweed host plant. Monarchs use these toxins to keep predators like birds at bay as the butterflies migrate thousands of miles from the eastern prairies of Canada and the United States to the subtropical mountains of Mexico. My father, like the butterflies, was using toxins co-opted from other organisms to keep his attackers, both psychological and physiological, at bay, but the toxins were just different ones. His long struggle, the tragic death spiral, and its subsequent impact on me serve as important touchstones throughout the book.

My attempt to grasp why he died allowed me to identify and then draw together the many ways that nature's toxins affect the world. So, my father's need for copious amounts of some of nature's toxins is really another thread, the most personal one, interwoven throughout. You may have had your own similar struggle or may have loved somebody with a substance use disorder. My hope is that your experience can be your own thread that you will weave throughout the book.

It was my father, who was a naturalist, who first taught me about nature's toxins. Not only did his knowledge rub off on me, but my growing up in northeastern Minnesota also had some bearing, as did my tendency to use nature as my own escape.

A child's morbid curiosity about species capable of biting, stinging, scratching, maiming, or poisoning is the bane of parents everywhere. There is always that one neighborhood kid, quiet but tempting fate, and I was that kid. Snakes that bite, toxic newts, snapping turtles, stinky burnet moths, stinging nettles that I couldn't stop touching (the itch needed to be scratched), and prickly porcupines that maimed our dogs—all these belligerent living things fascinated me. I recall like it was yesterday the confusion in my mother's eyes when as a kindergartner I presented her with a coffee can filled with a few hundred honeybees I'd collected as they visited the white clover growing in our neighborhood in Duluth, Minnesota.

Although I'd already had a few stings by then, I knew that the bees did so in self-defense. That was just the beginning of my intense curiosity about, and interest in, nature. I would wear garter snakes, which emitted repugnant cloacal secretions, draped around my neck. In Texas, I cupped horned lizards in my hands, enamored by how they could shoot blood from their eyes. In Nevada, I placed venomous black widows in salsa containers to bring home. I didn't inherit this love of nature and dangerous animals from my mom. My dad is the likely source. At that time, he was a used car salesman, and later, a furniture salesman, but in his heart, he was a naturalist.

When I was ten years old, we moved from Duluth to the Sax-Zim Bog, near the Minnesota townships of Toivola (the Finnish word for "hope"), Elmer, and Meadowlands. What I didn't know then was that the bog was a birder's paradise. It often hosts more great gray owls (*Strix nebulosa*) in the wintertime than does any other place in the continental United States, but few people live there. We moved out there, away from my mom's family in Duluth, so that my dad could take a better-paying job as manager of a (now closed) furniture store that had become a regional landmark. The store was in a fading farming community that drained the bogs to grow hay for dairy cows. Agriculture was successful in the wet meadows at the edges of the bog for a while, but by the time we got there, the population was aging and steadily declining in size. Most of the children had moved away to greener pastures as adults.

The local school, a single building for kindergarten through twelfth grade, enrolled around 150 students in total, including the 15 in my senior class in 1994. The school was shuttered a few years later. It had been one of ten schools in a district that was about the size of Connecticut and that stretched eighty miles, from Voyageurs National Park on the Ontario border down to the Sax-Zim Bog.

I was a closeted gay teenager and turned my energy toward the beauty of the bog, a few friends, and getting out of there. Nature provided a refuge and spiritual wellspring for me. And it continues to be a wellspring for me, both personally and professionally.

This book will also share insight into how evolutionary biologists like me tackle research questions. In that vein, note that I am only a biologist, not an anthropologist, a chemist, an ethnobotanist, a historian, or a social scientist. Nevertheless, the scope of this book is ambitious, and writing it required that I venture beyond the limits of my main areas of research. The roots of this book extend from my own life, through our recent past as a species, and into events buried by the sands of time, deep in evolutionary history.

Notes, including references used throughout, and an appendix containing further information on the toxins discussed are available online, through a link included at the end of the book.

1.

Deadly Daisies

Within the infant rind of this small flower
Poison hath residence, and medicine power.
— WILLIAM SHAKESPEARE, *ROMEO AND JULIET*

Rivers, Rings, and Reckonings

I pinned the boutonniere to my lapel, cataloging the species the florist had selected and the corresponding poisons in each. The star of the wintry bouquet was a bright, tiny chrysanthemum (mum) from the daisy family. It was surrounded by some needles of an eastern white pine, clusters of red berries of St. John's wort, and the spiky blue leaves of sea holly.

I hadn't requested poisonous plants on my wedding day, but I didn't have to. All plants produce chemicals that they can deploy as poisons to eliminate the competition, dissuade herbivores, neutralize pathogens, and punish unfaithful pollinators. Plants want to live, as do the many fungi, animals, and other organisms that also use poisons in offense and defense.

Even in its "infant rind," the mum carried a bevy of toxins, including the terpenoid matricin. The eastern white pine held its own piperidine alkaloids, St. John's wort contains the phenolic compound hypericin, and sea holly, the aldehyde eryngial.

You probably haven't heard of these chemicals, but each is also a medicine. Matricin is additionally found in chamomile and yarrow, plants used in traditional healing today and for thousands of years. In the body, matricin breaks down into the beautiful blue chemical chamazulene, which is now being studied for its promise as a pain-relieving drug. The needles of the eastern white pine have long been used by many northeastern Indigenous North American cultures to treat respiratory ailments. The piperidine alkaloids in the needles provide the starting point for the synthesis of opioids like fentanyl. Hypericin in St. John's wort is widely used to treat depression and other mental health disorders. Finally, Jamaican scientists discovered that the sea holly works as a traditional treatment for roundworm infections through the toxicity of the eryngial.

The big question is why plants would bother making these chemicals in the first place—after all, their synthesis takes up precious energy that could otherwise be put into growing and reproducing. One big hint came in 1964, when the late chemical ecologist Tom Eisner and collaborators published a paper showing that one species of millipedes produces eryngial (also called *trans*-2-dodecenal), the same chemical produced by the sea holly and other plants, including those in the citrus, ginger, and dill families.

The millipedes secrete eryngial when attacked by assailants like ants and grasshopper mice. The production of this substance in both animals and plants reveals a common pattern in evolution. The same beneficial trait often evolves in many organisms independently—in this case, eryngial as a defense for both animals and plants. The repeated origins of the same trait in different evolutionary lineages is called *convergent evolution*.

These natural toxins and their sources may sound more familiar than eryngial in millipedes. There is caffeine in coffee beans, cannabinoids in marijuana buds, capsaicin in red pepper flakes, cinnamaldehyde in cinnamon sticks, cocaine in coca leaves, codeine in cough syrup, and cyanide in apple seeds. It may surprise you to learn that many chemicals like these, which we use in food and drink, medicine, spiritual practice, recreation, and even for nefarious purposes like killing, are poisons produced by other organisms that did not evolve with us in mind. Yet these toxins permeate our lives in the most mundane and profound ways.

Such chemicals can be deployed as weapons in the Darwinian war of nature, which was first waged over four billion years ago, when life began. The battles in this chemical war continue to rage all around us, affecting the trajectory of each human life, including my own. Wherever we look, we find these skirmishes. For me, they are the markers of life and of death, the harbingers of joy and pain, and the vehicles of simple pleasures and wild rides.

As I began to write this book in rural Vermont, I also got married to Shane—in the dead of winter and on the solstice. We walked to the edge of the frozen, tea-stained river where Anne, a justice of the peace, awaited. As we trudged through the snow, I recalled a photo of my mother on her wedding day. In it, she holds a bouquet of oxeye daisies and stands on the bank of a blackwater river born in the boreal forest of Minnesota, much like the river Shane and I now stood above.

Downstream from the Lester River, where my mother and father, in matching lace outfits, were married, my father taught me how to fish in the dark eddies swirling below a waterfall that cut through the ancient basalt. When my four-year-old eyes beheld the first brook trout I pulled out of the rusty reach, the fish, like a miniature Georges Seurat painting, took my breath away. My father stood in front of me, smiling as I marveled at that living masterpiece of evolution. Ruby points with sapphire halos peppered the lower olive-green flanks, and neon green vermiculations were laid across its back.

That river transformed my father into a happier and calmer version of himself. But he couldn't take the river with him when he left. In the end, he died more than a thousand miles away from its waters and from all of us. In exile, he died alone, surrounded by an arsenal of guns and thousands of rounds of ammunition and hooked on what he called his "medicine." Periodic texts and calls to and from his flip phone were the only remaining threads connecting us.

On the morning of Christmas Day 2017, his lifeless, sixty-nine-year-old body was found by the county sheriff on the floor of a fifth-wheel trailer in West Texas. He had been dead for days, maybe even weeks.

By 2021, the box containing his ashes had been sitting for years in the same spot in our tiny house in Oakland, California, where I had first placed it—just below a shrine next to the window. The shrine held photos of him and me — snapshots of the arc of our time together. In August 2021, Shane and I put the box in the car with us and drove to Minnesota, en route to our sabbatical in Vermont. I just couldn't leave it behind.

Our last stop in Duluth was the home of my maternal aunt, who was also my godmother. She was on dialysis, owing to end-stage kidney disease, although she had received a lifesaving liver transplant about a decade earlier, after a long struggle with alcohol use disorder (AUD) had caused her own liver to fail. AUD is the clinical term now used in place of *alcoholism*.

I sat down on her sofa and felt her cold hand grip my forearm; her papery skin was held taut by the squeeze. She knew I was about to say something difficult, and she pulled me in close, staring into my eyes. I told her we were going to head to Vermont now, and on the way, I wanted to put my dad's ashes in the river. She said, "Yes, go down and do it, hon." "Hon" meant "honey," of course, but it was said in this particular northern Minnesotan way that pierced a thick shield around my heart. With that blessing, I hugged her tiny body. It was our last goodbye.

Just about a mile away from her house was the end of the river that had been such a large part of my father's life. It was finally time to let him go for good. We scattered his ashes at the mouth, right where it emptied into the "shining Big-Sea-Water" of Lake Superior.

Dark eddies enshrouded each flake of white bone, and then the current whisked the last pieces of him away, forever. Atoms of calcium and phosphorous, born in the heart of a star billions of years ago, could now continue their journey, diatom to mayfly to trout.

A few months later, Shane and I stood face-to-face at our wedding ceremony. My eye caught the chrysanthemum pinned just above his heart as the sun, slunk low like a blood orange in the winter solstice sky, illuminated a spiral of tiny petals. In the swirls of that small flower was an amalgam of the personal and the professional spheres of my life I had worked so hard to keep apart.

I saw the toxins in my own mother's bouquet, the substances flowing through the rivers, those that moved from plant to animal to me, those that took away so many in my family, and the chemicals central to my research.

They were all there, swirling together in that perfect spiral of the mum's petals.

Snowflakes spun in the cold, empty air. With our backs to the west, toward Minnesota, I slid a glittering ring of silver and gold onto Shane's finger. Onto mine he placed a ring of wood and amber, toxins entombed. I wouldn't have it any other way.

My ring was really three separate rings molded into one. The two outermost pieces are of black walnut wood and held juglone, a toxin that walnut trees make and that can kill competing plants that live under the trees. The wood also contained the dark tannins that strengthened the tree and that could deter most animals trying to eat it. The inner amber ring was fossilized resin of toxic terpenoids like alpha-pinene, produced by trees millions of years ago and used to defend against attackers.

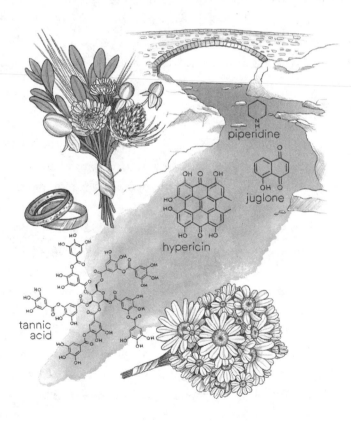

When I returned his ashes to the river that seemed to give my father new life each time he was near it, I couldn't help but think of another seemingly invincible man for whom a river was the source of his strength. For this man, too, it was a poison that eventually found his hidden vulnerability.

In the *Achilleid,* first-century Greco-Roman poet Publius Papinius Statius wrote of the goddess Thetis, who was forewarned of the death of her son Achilles. To thwart the plan, Thetis brought Achilles to the River Styx on the day of his birth. The waters of the river were supposed to confer to him the power of invulnerability. As she dipped the infant into the river, Thetis held Achilles by the heel, the one small part of his body that remained dry. Paris would one day exploit this vulnerability by driving a poison-tipped arrow right into Achilles' heel, mortally wounding him.

The injury-prone Achilles tendon reminds us that there is no foresight in evolution—no grand plan—or any plan at all, in fact. In truth, this tendon evolved from a much shorter and weaker one that serviced the hind foot of our tree-dwelling primate ancestors just fine, for tens of millions of years. These early primates lived in the trees and used all four feet and all twenty digits to grasp branches, just as many other primates do today.

As our own lineage began to transition from a life in the trees to one on terra firma, this tendon from the hind limbs of ancient arboreal primates was gradually repurposed by evolution into one used for bipedalism. Although the Achilles tendon works well enough for walking and running, it is far from an ideal solution to the problem of bipedalism, that is, the use of our hind legs as our only legs. Only a thin sheath and layer of skin separates it from injury, as anybody who has accidentally damaged their own knows.

Like Achilles himself, my seemingly invincible father was taken down by toxins that found a different and hidden vulnerability of ours: bodies that run on many of the same ancient chemical messengers and proteins as those in the animal enemies of plants, fungi, and microbes.

What I couldn't have known in the aftermath of his death was that my own research would not only provide comfort by distracting me from the grim situation but also help me understand the nature of his downfall. This book emerged from the collision of two worlds I once worked so hard to keep apart: my life's work to understand nature's toxins and my father's addiction to them.

I began to see how the fusion of these two parts of my identity could be useful in telling the story of nature's toxins. For example, the mums in the boutonnieres are more than just a metaphor for the most important events of my life. We can use them and other plants in the daisy family as a way to tease out the many concepts running through this book. We will start with a toxin that comes from some related daisies and that changed my own life.

Pyrethrum and Pests

It was a sunny spring day at the Saint Louis Zoo in 2001, and I was a twenty-five-year-old first-year PhD student in tropical biology fresh from receiving a master's degree in entomology. Strange as it may seem for a budding biologist who studied insects and plants, I was at the zoo to conduct a trial run of an experimental protocol I was planning to use on the wild birds of the Galápagos Islands. The research was led by my dissertation adviser, Patricia Parker, an ornithologist and a professor at the University of Missouri–St. Louis (UMSL). Our "guinea pig" was a resplendent red rooster. He was a perfect specimen, not a feather out of place.

While my friend and wildlife veterinary technician Jane carefully held him, I gently dusted his feather tracts with a natural flea and tick powder, hoping he wouldn't strike me with his spurs like I deserved. Then we waited.

Before I tell you exactly why I was doing this and what I was waiting for, you are probably wondering how this procedure is connected to daisies or nature's toxins. The insecticidal powder I used was made from the crushed, dried flowers of daisies classified as *Chrysanthemum* and, later, as *Pyrethrum*. Pyrethrum powder remains one of the safest and most widely used natural pesticides in the world.

When no synthetic ingredients are added, pyrethrum is appropriately labeled *organic, green, natural,* or *plant-based.* You have probably used pyrethrum at some point, whether it was in powder or spray form, on plants in the garden, on a pet, or on the living room carpet.

Make no mistake, pyrethrins, the active chemicals in pyrethrum powder, are incredibly potent neurotoxins—just not to humans. Human use of pyrethrins to kill typhus-carrying lice and plague-carrying fleas has long been a matter of life or death. Through trial and error, the Indigenous peoples of the Caucasus did the safety tests long ago.

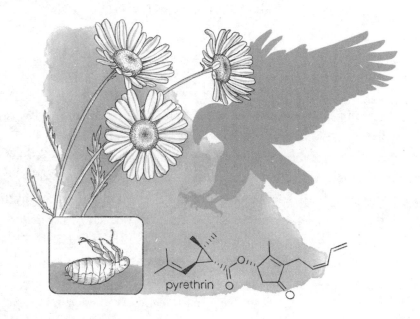

pyrethrin

Pyrethrum powder was first used as a pesticide in northern Iran, Armenia, and Georgia and was made from a different daisy species in the Balkans. "Persian powder" was introduced in Europe in the nineteenth century through an Armenian merchant. Then, in 1846, Johann Zacherl of Vienna mass-produced it from painted daisies grown in the country of Georgia.

In 1888, Zacherl's son built a factory in Vienna after switching to the feverfew daisy to make Zacherlin, the trade name for his formulation. Before 1888, the Russian military had also used pyrethrum powder to control fleas after learning of it from Circassian prisoners. Pyrethrins continue to be used in shampoos for the treatment of louse infestations—whether they are head, body, or pubic lice.

Chrysanthemums have been cultivated in East Asia for millennia, both as ornamentals and as medicinals. A testament to their antiquity in that region is the sixteen-petaled gold chrysanthemum on the seal of the emperor of Japan. The flower is an emblem of the Chrysanthemum Throne, the oldest hereditary monarchy in existence. However, pyrethrum-producing varieties of mums weren't grown in Japan until the early nineteenth century, when scientists there began to work on identifying the pesticidal chemicals in pyrethrum.

We now know how pyrethrum works as an insecticide: pyrethrins bind to important protein passageways (voltage-gated sodium channels) for sodium ions in nerve cells. When pyrethrins bind to these proteins, the nerve cells wildly overfire, causing involuntary muscle contractions, paralysis, and even death.

This physiological reaction sounds bad, and it is indeed problematic for invertebrates like insects (e.g., butterflies), mollusks (e.g., octopuses), arachnids (e.g., spiders), and some vertebrates, like fish. But natural pyrethrins are not very toxic to other vertebrates, like humans and birds. Dose for dose, the toxicity of table salt is higher than pyrethrins are for humans. The variable toxicity of pyrethrins in different species is due to the particular genetic changes found across the many branches of the evolutionary tree of life.

For example, a single ancient change in the DNA of insects makes their nerve cells a hundred times more sensitive to pyrethrins than ours. By contrast, cats and fish are sensitive to pyrethrins because they lack one of the liver enzymes we humans use to detoxify pyrethrins.

However, other natural toxins targeting the very same nerve channels are poisonous to us, even in tiny doses. Consider the sad story of the twenty-nine-year-old Oregon man who swallowed a rough-skinned newt on a dare. Just ten minutes later, his lips began to tingle, and in a few hours, he was dead. Tetrodotoxin in the newt's skin killed him. Like pyrethrins, tetrodotoxin targets the voltage-gated sodium channels but does so in a different location on the protein.

Although pyrethrins are made by plants, tetrodotoxin is made by symbiotic bacteria living in some freshwater and marine animals, including puffer fish, blue-ringed octopuses, and newts. The toxin is not produced by the animals themselves. In turn, the toxin-adapted, voltage-gated sodium channels in these animals renders them completely resistant to the tetrodotoxin.

Just as we and some daisies use pyrethrins as tools to keep disease-causing pests at bay, these animals deploy tetrodotoxin as toxic defenses against attack. Although we are susceptible to tetrodotoxin at small doses but highly resistant to pyrethrins, the puffer fish, blue-ringed octopuses, and newts are highly susceptible to pyrethrins. The selective toxicity of pyrethrins reveals why they, but not other voltage-gated sodium channel toxins like tetrodotoxin, are safely used for many applications, from mosquito control to flea powder and, as we will soon see, for removing lice from endangered birds.

The lesson? Pick your poison—carefully. The origin story of each of these chemicals holds critical information on why the benefits may or may not outweigh the costs for human use. There is nothing inherently healthy about natural products.

Nature, Red in Tooth and Claw

The first time I laid eyes on a Galápagos hawk, I was intrigued by how much it looked like the hawks I'd seen in northeastern Minnesota. But looks can be deceiving.

An adult female nearly knocked me unconscious after swooping down from on high and raked her talons across my face as a parting gift. My lab mate, who was the lookout for the hawks defending their territories from interlopers like us, just hadn't seen her coming in. It was an unfortunate accident.

That Galápagos hawk was simply doing what many birds—be they tree swallows or golden eagles—do if a human gets too close to their nest. They dive at the intruder's head.

The bird was completely unharmed in the aftermath, but when I opened my eyes, I could only see out of the left eye. Fearing she would come after us again, we ran under a spiny acacia to catch our breath as blood dripped from my eye, nose, and cheeks. Poisons aren't the only weapons used in the war of nature.

Thankfully, I was only temporarily blinded. Blood from a punctured eyelid was only what was dripping into my eye.

But the relief was short-lived. To our astonishment, the hawk circled back, landed on the ground just a few feet away, and moved toward us. Despite the hilarity of her pigeon-toed walk, we feared she was actually preparing for a ground assault.

Many of the animals in the Galápagos evolved in the absence of human persecution. As a result, they tend to be fearless. Darwin remarked that the birds, including the hawks, on the islands were so unafraid that "a gun here is almost superfluous; for with the muzzle of one I pushed a hawk off the branch of a tree."

In my mind, we had unwittingly re-created the raptors-in-the-kitchen scene from *Jurassic Park*. In that scene, two frightened children hide behind the kitchen counter to escape the velociraptors stalking them.

Although pint-sized compared to a velociraptor, hawks are the top land predators in the Galápagos. I once observed a hawk perch above a pregnant goat obscured under the cassia and grasses. The goat was attempting to hide and prevent her impending newborn kid from the hawk's skull-crushing claws. The hawk patiently watched and waited as the goat went into labor.

If you've been to the Galápagos or watched nature documentaries filmed there, the scene I described is a typical one. Fear, suffering, and death are everywhere, as the war of nature plays out in broad daylight.

In contrast, the chemical battles raging between species, between the poisoners and the poisoned, are largely hidden from our view. Once the veil is lifted, however, their impact on evolution, our daily lives, and even our recent history as a species is far more pervasive and dramatic than you could have known.

Later that night, after my brush with the hawk, I licked my wounds back in my tent on Isla Santiago. I was camped with our Ecuadorian collaborators on the sunbaked mud of a dried-up lagoon behind Espumilla

Beach in James Bay. Under the glow of a flashlight, I was reading *The Voyage of the Beagle*, which my adviser had handed to me shortly before I left St. Louis for Ecuador. I learned that Charles Darwin and I were the same age, twenty-six, when he visited that same beach in 1835.

I tried to close my eyes, but the moonlight was too bright. Then I heard the muffled sounds of hatchling green sea turtles moving through the sand on their way down to the bay. They traveled at night because that's when their principal enemies were sleeping, beaks tucked under wing.

However, other predators did emerge in the inky darkness of the Galápagos night in search of prey like hatchling turtles. When I lay back down, I made out the silhouette of the venomous Darwin's goliath centipede, all twelve inches of it, resting on the top of my tent. One of the largest in the world, this centipede attacks small mammals and birds, subduing them with a potent cocktail of protein-based venom injected through scythe-like mouthparts.

It was thrilling to be surrounded by such vibrant life, each organism doing its best to survive. I felt completely at home even though we had no satellite phones and only a marine radio in case we needed to call for help. The extreme isolation, desolation, and potential danger were comforting to me. I attribute this sense of ease to where, and how, I was raised.

Before working in the Galápagos, I knew from the scientific literature that the hawks harbored their own feather lice, or *piojos,* as our Ecuadorian collaborators called them. For my PhD research, my idea was to do something that Darwin couldn't have imagined.

I wanted to use mutations that had naturally accumulated in the DNA of the hawk's feather lice as an evolutionary tracer of the hawk's colonization history as this species island-hopped over hundreds of thousands of years. It was likely that each hawk got its lice from its mother, meaning the lice were inherited from generation to generation, much like the hawk's own genes—as an unwanted heirloom. To test this idea, I had to first get the lice off the hawks.

I did have some experience in that department. When I was twelve years old, I shot my first ruffed grouse, an upland game bird, in the forest behind our house. When I set the lifeless bird on the hood of our car back at the house, countless feather lice, as white as snow, crawled onto the dark metal. They jumped ship because the bird's body had cooled, and the minute creatures had mistaken the warmth of the hood for a potential host.

Of course I couldn't have known that I'd one day actually study birds and their lice as research subjects, but that boyhood event helped prepare me. In the Galápagos, we needed to get the lice off the birds without harming the birds in any way. Fortunately, the pyrethrum powder I dusted on the rooster in St. Louis as a test case was approved for us to use by the Galápagos National Park for our research because it was harmless to the hawks. The powder worked like a charm. The first hawk that I "dust-ruffled" with pyrethrin dropped hundreds of lice onto the plastic tray I borrowed from the Charles Darwin Research Station cafeteria. Through our research, we learned that the DNA of the lice could be used to retrace the hawk's colonization history in the Galápagos.

Pyrethrum is used for a more practical purpose on the islands. The mangrove finch, an endangered species of Darwin's finch, lives on just a few islands in the smallest populations. It is being ravaged by an invasive, parasitic vampire fly whose larvae feed on the blood and facial tissue of the nestlings, often killing them in the process.

To control these awful flies, park managers have attempted to treat finch nests with pyrethrins, which are toxic to the vampire flies but not the birds. But finding and treating nests is difficult work and all but impossible on a large scale.

Cleverly, ecologists Sarah Knutie and collaborators took advantage of the fact that Darwin's finches collect the soft cotton fibers from wild Darwin's cotton plants to line their nests. The scientists strategically distributed store-bought cotton treated with pyrethrins throughout the nesting sites of Darwin's finches near the research station. Sure enough, the birds gathered the insecticide-laced cotton and lined their nests with it. The birds that did use the treated cotton reduced the vampire fly infestations in their nests, thereby increasing the odds of nestling survival.

As staged as Knutie's experiment was, one animal does much the same thing in nature. Russet sparrows in China gather the leaves of wormwood, which somehow repels or reduces parasites, to line their nests. This behavior naturally increases survival of nestlings.

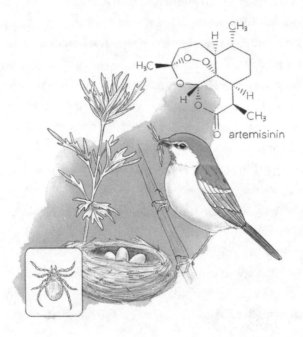

Wormwood is another member of the daisy family, and although its name may seem exotic, it is a close relative of the sagebrush that typifies the landscapes of the American West in both sight and smell.

The russet sparrows aren't alone in their use of wormwood to ward off enemies. Humans use wormwood too. When the birds start to nest in the spring, it's also the time of the Dragon Boat Festival. Because this holiday coincides with the beginning of warm weather, tradition calls for the hanging of wormwood sprigs from doorways to ward off pests. As these examples show, human and animal behavior often mirror each other when it comes to using nature's toxins as tools.

Wormwood contains a plethora of toxins. One is a terpenoid called artemisinin, which was discovered by physician Tu Youyou in China in 1972 as a treatment for malaria. Her research landed her the Nobel Prize in Physiology or Medicine in 2015, and artemisinin remains the gold standard for an antimalarial drug.

Despite our commonalities with other animals, there is an important difference between them and us with respect to self-medication. Our use of nature's toxins as tools is probably mostly learned. Other animals' use of toxins, on the other hand, seems to be mostly innate.

The use of toxic tools seamlessly evolved into the oral and written healing traditions I will collectively call the *materia medica*. These substances are found in every culture. In fact, those first used by Indigenous healers have yielded nearly 50 percent of all modern drugs we use today.

Things get fuzzy, and most interesting, when we look at our primate relatives as well as our primate ancestors and their use of nature's toxins as medicines. Self-medication in the animal world, or *zoopharmacognosy*, has evolved in many species, including, of course, our own. When describing human use of nature's toxins, though, we drop the "zoo" and call it *pharmacognosy*, which means "knowledge of drugs." We like to think we are special—and we are, in many ways—but in other ways, we are not. The truth is that what we like to call modern medicine is just the culmination of millions of years of pharmacognosy in our own primate lineage.

Profens and Paleomedicine

Human use of pyrethrin daisies and wormwood is ancient, but just how ancient? Three other members of the daisy family show us just how far by providing an even older link between our modern lives, our ancestors, and the origin of the pharmacopoeia. The three plants are yarrow, chamomile, and one called *Vernonia*.

In the *Iliad*, Homer describes how Achilles' army, through the advice of his teacher, the centaur Chiron, was instructed to carry yarrow to treat injuries. In real life, the ancient Romans called it *herba militaris*, and later, yarrows were given the name *Achillea*. Yarrow is known colloquially as bloodwort or nosebleed for its anticoagulant effects. In the 2,000-year-old

De Materia Medica, the Greek botanist-physician Pedanius Dioscorides promoted the plant's use both for wound healing and to combat dysentery.

In addition to expounding on the virtues of yarrow, British physician Nicholas Culpeper's *Complete Herbal* from the 1600s explains that chamomile, another daisy, "expels wind, helps belchings, and potently provokes the menses: used in baths, it helps pains in the sides, gripings and gnawings in the belly." Sounds about right! I love my sleep-inducing chamomile tea just before bed.

Yarrow, chamomile, and other daisies like wormwood were key medicinal plants and remain widely used around the world as medicines. In our own "poison garden" in our backyard in Oakland, Shane planted below our lemon tree a striking variety of yarrow with blood-red flowers. Every now and then I pluck a few sprigs of the yarrow and inhale just to experience the intense, unforgettable smell of the plant, which produces volatile chemicals like the phenolic matricin.

Few herbivores can overcome yarrow's bitterness. Matricin is one reason why. This chemical is in the terpenoid class and is also produced by other daisies like chamomile and wormwood. Matricin breaks down in our bodies into chamazulene, which I mentioned earlier in this chapter, and produces that beautiful blue color when dissolved. But it has another virtue: it is a profen, a drug that reduces inflammation. Other profens include the nonsteroidal anti-inflammatory ibuprofen. Chamazulene may work in the same way as ibuprofen in our bodies. So, chamomile could be more than just the key ingredient in sleep-promoting chamomile teas.

The ancients were onto something. But on their own, these anecdotes take us no further than a few thousand years into the past. When I mentioned that some members of the daisy family can provide a link between our ancestors and the origin of the pharmacopoeia, I was referring to more distant human ancestors—the Neanderthals, *Homo neanderthalensis,* a species that went extinct around thirty thousand years ago.

Amazingly, Neanderthals live on in billions of people, because many of us carry stretches of Neanderthal DNA in our chromosomes. The reason is that our *Homo sapiens* ancestors interbred with *Homo neanderthalensis* when the two groups met outside Africa.

The more ancient African *Homo* lineage that included *H. neanderthalensis* eventually settled in Europe, the Middle East, and western Asia well before *H. sapiens* left Africa. Eventually, *H. neanderthalensis* went extinct and only *H. sapiens* remained outside Africa. However, because of this interspecies hanky-panky, the genetic legacy of Neanderthals lives on in many of us.

In addition to their DNA, the nature of Neanderthals is chronicled through artifacts and bones left behind in caves. Based on their cranial sizes, they certainly had brains as big as ours, if not slightly larger.

Neanderthals also probably shared our ability to taste bitter chemicals. People vary in their like or dislike of certain bitter plants such as broccoli and brussels sprouts. This difference is in part determined by the genetic variants a person carries for a gene called *TAS2R38*, which is expressed in our taste buds.

The common form of this gene allows people to taste bitter chemicals like phenylthiocarbamide (PTC), which has long been used by scientists to screen individuals for bitter detection. Those of us carrying at least one copy of this common *TAS2R38* gene variant, which is dominant, are called PTC tasters. Those of us carrying two copies of the less common form of *TAS2R38*, which is recessive, are called PTC nontasters.

The bones of seven Neanderthal adults and six Neanderthal children, who died around fifty thousand years ago, were found in El Sidrón, a cave in northwestern Spain. The genome sequences obtained from a tiny bone flake of one of the adult males revealed that he was a PTC taster. This matters for what was discovered next.

Scientists scraped the calcified tartar, or dental calculus, from the ancient teeth of several Neanderthals from El Sidrón. If you roll your tongue across the inside of your lower two front teeth, you may feel a bit of your own dental calculus.

Fortunately for us, Neanderthals didn't have dentists. Important evidence of how they lived their lives was left behind, deep in this calcified tartar. Sophisticated chemical analyses of the tartar identified some of the plants these individuals were eating and what chemicals were in the smoke they inhaled. DNA sequencing methods even divined the identities of the long-dead microbes once living in their mouths and the plants they consumed—plants that couldn't be identified through the chemical analyses.

One Neanderthal individual, known as El Sidrón Adult 1, or more affectionately as "Sid," was probably quite ill when he died. The scientists suspected this condition because Sid had a dental abscess and was infected with a microsporidian pathogen that causes a diarrheal disease. (Poor Sid!).

Sid's tartar contained traces of toxins from yarrow and chamomile plants, including chamazulene, as well as DNA sequences from poplar trees and the fungus *Penicillium*. Poplar trees are a traditional source of salicylic acid, a type of salicylate used to make aspirin. *Penicillium* species are molds that produce the antibiotic penicillin. Sid may have been treating himself for his ailments, or he may have been treated by another Neanderthal, using the same medicinals we use today, some fifty thousand years later.

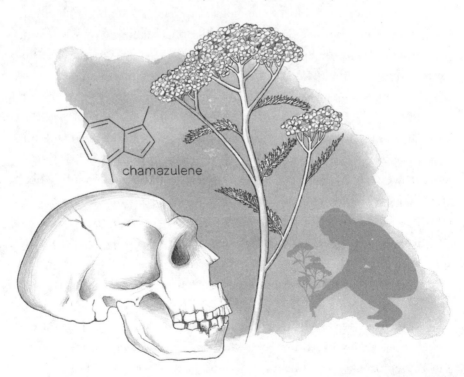

chamazulene

Of course, we will never know if Sid purposefully consumed these materials to treat his ailments, but yarrow, chamomile, poplar, and *Penicil-*

lium do not have nutritional value. Because at least some Neanderthals like Sid could detect the bitter chemicals in each of these sources, he would probably have perceived these substances the same way that most of us would perceive them: distasteful.

So, perhaps Sid knew what he was doing, and despite the bitter taste, the benefits of consuming these items was worth the unpleasant flavor. This hypothesized behavior is not much different from the way I overcame my aversion to the bitter chemicals in coffee in exchange for a caffeine buzz. We will look at caffeine in more detail later. Exploiting nature's toxins for our own benefit, despite some clear downsides, is an essential part of what it means to be human.

As amazing as the chemical secrets Sid's tartar held for fifty thousand years are, a sample size of one, although better than none, may be as good as it gets when it comes to Neanderthals and evidence of early self-medication in ancient *Homo*. Fortunately, we can tap into another evolutionary time machine. Of the primates living today, the common chimpanzee and the bonobo (also called the pygmy chimpanzee) are our closest relatives. We diverged from a most recent common ancestor with them around five to ten million years ago.

Tantalizing data from observations of common chimps in the wild suggests that they self-medicate using plants in the daisy family too. These observations further support the notion that self-medication is all in the family, and although self-medication seems to be an essential element of human behavior, it isn't exclusive to us.

One of Shane's favorite dishes when he lived in Cameroon as a Peace Corps volunteer was *ndolé*, a West African stew made of vegetables and meat. Leaves of a plant called bitter leaf are a critical ingredient. Because bitter leaf comes from various species of equatorial African bushes in the tropical daisy genus *Vernonia* and is hard to find in the United States, we can't quite re-create the recipe using ingredients in California.

Vernonia plants are relied on heavily as medicinals across West Africa. People use them to treat many diseases, particularly infectious diseases like malaria, schistosomiasis, and intestinal parasites.

In 1989, traditional healer Mohamedi Seifu Kalunde and his

collaborator and primate researcher Michael Huffman were in Mahale Mountains National Park of Tanzania. They observed an obviously sick adult female common chimpanzee over two days. She repeatedly sought out shoots of wild *Vernonia,* carefully removed the leaves and outer bark, chewed on only the inner pith, swallowed its bitter juice, and spat out the remains. Like Sid, she was the sole chimp observed in her group to exhibit this behavior, and unlike us, chimps do not normally use the plant as a source of nutrition — they can't make ndolé even when they have access to *Vernonia*!

Over the next decade, the researchers observed many other chimps in this population chewing the bitter pith. The scientists eventually hypothesized why the chimps were doing this. The use of bitter leaf was associated with an uptick in intestinal nematode infections in the chimp population.

Reductions in the chimp's worm burden — in other words, recovery from worm infection — were observed after bitter pith chewing, suggesting that the plant's toxin works as a dewormer. Like Sid, the chimps seemed to know what they were doing. Not only is food eaten for sustenance, but it may be ingested for medicinal purposes, too.

Humans and chimps aren't the only primates to self-medicate. Gorillas and orangutans, our next-closest relatives outside of chimps, do it, too. Remarkably, just as in the case of *Vernonia,* the adjacent Indigenous human populations use these *same* plants as medicines. For example, in Borneo, ten orangutans were observed to chew the leaves of *Dracaena* plants into a soapy froth that the animals then applied to their fur to repel parasites or treat skin diseases. Indigenous people living in the same forests that the orangutans occupy also use a poultice of leaves from this plant to treat several maladies.

The main toxic constituent of the *Dracaena* leaves are saponins, bitter steroidal chemicals made by many plants as defenses against enemies. We use some of these plants, like soapworts from Europe and Asia and soapbark trees from South America, to make soap. Remarkably, a saponin from a soapbark tree native to Chile is used as an adjuvant in the Novavax vaccine that the US Food and Drug Administration (FDA) has approved to help prevent COVID-19 disease.

These examples suggest that humans have repeatedly used the know-how of our closest relatives in pursuit of the healing powers of plants and other organisms. Or perhaps it's the other way around: these hominid relatives might be watching and learning from us. Maybe it is a bit of both.

In this chapter, we've used daisies as a microcosm to examine the origins of nature's toxins, how both humans and other animals use these toxins as tools, and how toxins can change the course of our own lives. As we have seen, plants in the daisy family produce a diversity of toxin classes, including alkaloids, flavonoids, phenolics, and terpenoids.

The sheer number of toxins made by just one plant family reveals the staggering diversity of nature's toxins that can affect us, often unwittingly. However, there were other plants in my boutonniere, and I couldn't avoid bringing toxins from other organisms besides daisies into the story. So, we now need a different approach.

Instead of surveying one class of organisms at a time, I will now focus each chapter on just one or two broad toxin classes at a time, just as a chemist would. Let's first examine some of the most ancient and diverse chemicals: phenolics and flavonoids.

2.

Forests of Phenolics and Flavonoids

No way was clear, no light unbroken, in the forest. Into wind,
water, sunlight, starlight, there always entered leaf and branch,
bole and root, the shadowy, the complex.
   —URSULA K. LE GUIN, *THE WORD FOR WORLD IS FOREST*

The Story of Tannins, Part 1: White Wine, Black Water, Purple Ice, and Green Tea

My brother and I used to compare the water flowing in the Lester River to root beer. We were onto something. The chemicals leaching from the boreal forests and bogs that the river drained have the same foaming and tinting effect that they do in root beer, which is traditionally made from sarsaparilla roots.

Although I have a special fondness for blackwater rivers that flow into Lake Superior, there are far more impressive and ecologically and economically important ones in the tropics. The Rio Negro, one of the largest tributaries of the Amazon River, along with the Congo River, are the most well known. Whether these rivers flow through a boreal forest or a tropical rain forest, they all drain densely vegetated lands that produce enormous quantities of tannins and other phenolic and flavonoid compounds responsible for the rivers' tea-stained color.

Rivers and my wedding ring aren't the only places tannins have shown up in my life. My favorite scene in the movie *The Birdcage* occurs just before

Albert, played by Nathan Lane, goes onstage as his drag character Starina. Albert accuses Armand, played by Robin Williams, of having an affair.

Albert found a bottle of white wine in the refrigerator, but Armand drinks only red. Armand's explanation was that he was switching to drinking white because red has tannins. The alibi apparently worked because red wines *do* tend to have more tannins than white wines. Nonetheless, some white wines, like chardonnays, are highly tannic—the oak barrels in which chardonnays are aged leach tannins from the wood into the wine.

I avoid barrel-aged wines like these because I am sensitive to oak-derived tannins. Sparkling wines are not aged in oak barrels, so they are my go-to when a drink is in order. Nonetheless, while I may dislike oaky wines, I am mesmerized by the beauty of the chemical structures of tannins.

The individual gallic acid molecules bound to a central glucose molecule in tannic acid make a perfect pinwheel. Although the structure is satisfying to the eye, it takes precious energy for a plant to make each of the chemical bonds holding these carbon, hydrogen, and oxygen atoms together—energy that could be put toward growth and reproduction. In the economy of nature, tannins aren't cheap.

Tracing the origins of tannins will lead us to a rather startling conclusion that bears on the evolution of land plants, the Industrial Revolution, the founding of the United States, and even the melting of the Greenland ice sheet. To get there, we will need to wade into the chemical weeds, but just for a bit.

Tannins are made of either phenolic or flavonoid molecules, but all tannins are defined by their ability to bind to proteins. Because tannins are so good at binding to protein, they have long been used to tan the hides of animals in the making of leather. The tanoak tree of the Pacific coast of North America shares an etymological root with the older terms tan and tannin.

You may have heard of phenolics and flavonoids before, as well as the term polyphenols, which describes molecules made of several phenolic or flavonoid molecules bonded together. These chemicals are often called antioxidants because they mitigate oxygen radicals, by-products of our body's normal metabolism. Oxygen radicals, chemically unstable forms of oxygen, can bind quickly with other substances and injure healthy cells.

For now, I will restrict this discussion to two main classes of tannins made by plants: hydrolyzable tannins and condensed tannins. Hydrolyzable tannins are derived from phenolic molecules made by an ancient metabolic pathway found in bacteria, algae, plants, and fungi but not in animals. Condensed tannins are flavonoids made by a more recently evolved pathway restricted to plants.

Hydrolyzable tannins are further classified into two types, gallotannins and ellagitannins. Remarkably, gallotannins and other hydrolyzable tannins have also been found in the closest living relatives of plants, green algae in the family Zygnematophyceae, whose members live in both fresh water and on land. Gallotannins are one answer to how these algae can live on the surfaces of glaciers and mountaintops.

Ultraviolet B (UVB) light is highly damaging to DNA and other molecules of life. UVB light rapidly dissipates under water, so aquatic algae are more shielded from it than those growing on land. The planet's ozone layer offers some protection, but not quite enough.

When Zygnematophyceae algae are exposed to sunlight on land, they turn purple. The purple pigments are gallic acid molecules that form a complex with iron. This pigment absorbs UVB light from the sun and reflects purple light back into the environment. In this way, gallotannins help prevent DNA damage caused by UVB radiation from the sun.

When these algae grow on ice, the creation of their own sunscreen turns the ice gray or purple. In Greenland, where Zygnematophyceae algae colonize its vast ice sheet, the hydrolyzable tannins are dark; they absorb heat from the sun's rays, increasing the rate of melting. As a result, these tannins from algae are exacerbating the sea level rise caused by greenhouse-gas-induced global warming.

Hydrolyzable tannins probably first evolved as sunscreens. Once plants evolved on land, these compounds took on additional roles, such as discouraging consumers.

Condensed tannins evolved more recently than hydrolyzable tannins and are found only in plants, not green algae. Made of two or more catechin molecules, condensed tannins are found in abundance in tea, acai berries, apples, cinnamon, cocoa, grapes, and oak.

In moderate amounts, condensed tannins can be beneficial to our health and wellness, but they do serious liver damage when consumed in large quantities. Around 10 percent of green tea extract consists of condensed tannins. The most abundant one, epigallocatechin-3-gallate (EGCG), shows some promise in protecting against a variety of diseases, from diabetes to dementia, at low doses. Infusions of green tea in hot water or food, like those made from matcha, my favorite green tea powder, are generally safe.

However, EGCG pills have become a popular weight loss supplement, and this is where the old adage from Paracelsus, "The dose makes the poison," comes into play. When EGCG doses in supplements exceed eight hundred milligrams per day for sustained periods (one to six months), liver toxicity can occur. Reports of liver injuries associated with high doses of EGCG were so concerning that the European Food Safety Authority issued a rare public warning in 2018.

I became aware of the dangers of condensed tannins after reading the stories of well over one hundred people (some as young as sixteen years old) who have suffered acute liver failure in association with green tea extract supplements. The US National Institutes of Health (NIH) even published an online book for physicians and researchers. *LiverTox* focuses on toxins found in the highly unregulated herbal and dietary supplement industry. Buyer beware.

Plants can also use a precursor of condensed tannins called cinnamic acid to make a variety of other chemicals, some of which are used as defenses and others for normal growth and tissue maintenance. Stilbenoids are one of these other cinnamic-acid-derived chemicals and include resveratrol from grapes that gained (and then lost) fame as an antiaging drug. Another derivative of cinnamic acid is cinnamaldehyde, which gives cinnamon its characteristic odor and flavor.

Coumarin, which underlies the pleasant smell of sweetgrass, is another molecule naturally synthesized from cinnamic acid. Coumarin played an important role in the development of modern medicine. A modified form of this compound was the inspiration for the synthesis of a rat poison and the anticoagulant warfarin, sold as the blood-thinning drug Coumadin. Yes, a rat poison is also a medicine.

The story of coumarin and warfarin (brand name Coumadin) is fascinating. In the 1920s, a mysterious disease caused cattle to bleed to death shortly after they had been castrated or dehorned. Besides the surgeries, the cattle had something else in common. Each had eaten sweet clover that had been inadvertently infected with a particular kind of mold. Researchers connected the dots and discovered that the fungus had chemically transformed the coumarin made by the sweet clover into a different chemical, dicumarol. This chemical caused the uncontrollable bleeding because it interfered with the cow's prothrombin, a factor necessary for clot formation.

The chemical structure of dicumarol was eventually used as the basis for the synthesis of new drugs with similar effect, including the synthetic drug warfarin. Warfarin is also effective as a rodenticide but moves up the food chain, killing predators like eagles, owls, and mountain lions, and its use is banned in British Columbia because of these nontarget effects.

Beyond their use in drink, food, and medicine, people have also used these tannins to manufacture leather and even to create some of the most important historical documents ever written, including the Dead Sea Scrolls and the Magna Carta. Let's look at these practices next.

The Story of Tannins, Part 2: Perfumed Gloves and Iron Gall Ink

As a junior in college, I participated in a study abroad semester in France, along the Côte d'Azur. Our dorms, although not quite on the Promenade de la Croisette, were close enough. Voilà, we lived right across the street from the beach.

The evening my classmates and I arrived, I ran into the crystalline turquoise water with them and simply couldn't believe I was on the other side of the planet, swimming in the August waters of la Méditerranée. The water trickling down my face hid tears of joy. My most fervent desire was to travel and learn as much as I could about whatever was before me. I took daily, difficult immersion courses in French, a course in art history, and one in Roman culture and history.

The Roman course was taught by the Reverend Jerome Tupa, a Roman Catholic priest, Benedictine monk, artist, and French professor. Using his connections and knowledge, Father Tupa took us on weekend excursions to places like the abbey of Saint-Martin-du-Canigou. Built in 1009 CE, the abbey was perched high on the crest of a mountain in the Pyrenees and was run by a Catholic group called the Community of the Beatitudes.

During our visit to Saint-Martin-du-Canigou, we stayed at a different abbey, which was run by aged, white-bearded Benedictine monks who prepared dinner for us and, of course, provided wine they had made. After each drink, the wine puckered my lips and dried my mouth, thanks to the astringent properties of the tannins found in the skin and seeds of the Catalonian grapes used to make it. In our mouths, and those of other mammals, tannins bind to salivary proteins that have an affinity for these chemicals — in fact our salivary glands immediately begin to produce these proteins when we ingest tannins. When the tannins bind to the salivary proteins, the tannins precipitate out of solution in our saliva, creating the rough, dry puckering sensation that many of us grow to like.

The levels of tannins found in alcoholic drinks are harmless to us humans, but that isn't necessarily so for other animals that encounter tannins in nature.

Inside the human mouth, the salivary proteins bind with tannins. By mopping up the tannins, these proteins prevent them from wreaking havoc in our digestive tracts later on. In this way, salivary proteins lower the dose of intact tannins that enter our digestive tracts. Left to their own devices, some tannins can bind to other proteins in our intestinal cells and thereby prevent us from absorbing nutrients. Tannins can benefit plants because they can also harm pathogenic microbes and herbivores by disrupting the ability to absorb nutrients.

Tannic and gallic acids were once used in enemas. But like drinking excessive amounts of green tea catechins, the excessive use of these types of enemas in the mid-twentieth century caused acute and fatal liver failure in eight people in the United States. For this reason, the use of hydrolyzable tannins in enemas was banned in 1964.

Our relationship with tannins extends beyond food, medicine, and accidental poison. To examine it, we'll have to learn a bit about one of the most obscure organisms on the planet: the oak gall wasp. If you live around oak trees, you have probably seen odd objects the size of ping-pong balls on the ground beneath the trees or even on the leaves or twigs above you. These are oak galls, or oak apples, which are hard on the outside and filled with a soft, pillowy tissue, which you can see if you break one open.

The galls are neoplasms, or new tissues, similar to cancers in animals. Like some cervical and throat cancers that are caused by the human papillomavirus, galls are also caused by an infection. In oaks, the galls are induced when a wasp in the family Cynipidae injects venom and an egg into the base of a leaf bud. Through largely unknown mechanisms, the plant's hormones are co-opted by the wasp venom to produce both shelter and food for the grub that hatches from the wasp's egg.

As the oak gall grows, it accumulates tannins. The oak trees produce tannins that protect their leaves from pathogenic microbes and herbivores. But in the gall, the wasp's venom hijacks the oak's tannin production machinery. As a result, the gall is just loaded with tannins, which may protect the wasp's grubs from herbivore and fungal competitors as well as insect pathogens.

Remarkably then, the tannins in oak apples or galls may be a by-product of the struggle between plants, wasps, and fungi. This dynamic played an essential role in the formation of my own country, the United States. I saw this connection firsthand as a ninth-grader, when I participated in the Close Up program in Washington, D.C., which showcased the workings of the federal government to high school students.

While visiting the capital, we students eventually made our way to the National Archives to see the Declaration of Independence. I was taught to believe this eighteenth-century document was a sacred one because of its connection to the modern conception of universal human rights. We stood in hushed silence and made out the large-lettered preamble: "We hold these truths to be self-evident, that all men are created equal." The words gave us goose bumps.

I had flown to D.C. with my high school guidance counselor and two other students from Orr, Minnesota, eighty miles to the north of where I lived. After our trip to Monticello with my new friends, who happened to be members of the Bois Forte Band of Chippewa, we felt conflicted. We tried to reconcile how the self-evident "truths" embedded in that document were written by a man who had enslaved over six hundred Black people and referred to Native Americans as "merciless Indian savages." The virtue preached in that document could not be separated from the vice practiced. The issue was complex. There was a duality we couldn't fully parse.

What I didn't know then was that however transcendent and contradictory its words and impact were and are, the Declaration of Independence and my country's other founding documents were created using iron gall ink derived from the tannins in oak galls made by those tiny wasps.

Roughly between 400 and 1800 CE, Europe and its colonies relied nearly exclusively on oak galls as the source of ink for writing. Even more notable, gall ink without iron was likely used seven hundred years earlier by the calligraphers of the Dead Sea Scrolls. These ancient manuscripts captured the imagination of the world when they were discovered in 1948, because they represented the oldest texts that would become part of the biblical canons.

When hydrolyzable tannins in the oak gall are mixed with iron salts, they form a new compound with a blackish-blue hue, the iron gall ink. This chemical reaction between iron salts and tannins is the same one that gives the algae living on glaciers their purple hue and is similar to that found in some mushrooms that turn blue after they are wounded. We will look at some of these mushrooms in more depth later.

Iron gall ink and quill pens were also used to write the Magna Carta, which may have helped inspire the Declaration of Independence. Leonardo da Vinci (1452–1519) also used this ink in his drawings. Unfortunately, iron gall ink fades with time and eats away the very paper it adorns, posing major challenges for the preservation of these documents of incalculable value.

tannic acid

Shortly after arriving in Cannes during my study abroad, we visited three perfumeries in the Grasse prefecture on the Côte d'Azur: Fragonard, Galimard, and Molinard. These perfumeries trace their origins to Catherine de' Medici (1519–1589), through tannins. I recall rooms with long tables crowded with giant bowls of dried flowers and jasmine-infused air. Although Grasse is synonymous with perfume now, it wasn't always that way. The perfumeries there have de' Medici and tannins to thank for their success.

For hundreds of years, starting in the eleventh century, long before it became the perfume capital of the world, Grasse was *the* leather-making city in Europe. The area supplied high-grade leather to northern Italian towns that specialized in making products like gloves, belts, hats, and footwear. The bark of oaks, pines, and other local plants was used to preserve animal skins.

The perfume industry for which Grasse became famous can be traced to Catherine de' Medici's arrival in France. Italian by birth, she was queen consort to France's King Henry II. She fancied the perfumed gloves widely available in Italy and was, according to legend, given a pair by a tanner named Galimard from Grasse.

Scents were made from a pomade of lard imbued with extracts of flowers and spices. The scented lard was then used to line the leather gloves. For one thing, the perfume hid the awful smell of the leather. And second, a wearer, by bringing the gloves to the nose, could also overpower any fetid odor encountered in the city. Such odors were not uncommon at the time, even at the Château de Fontainebleau. They were erroneously thought to be the source of infectious diseases like the plague.

Catherine de' Medici's trendsetting gloves sparked high demand across Europe. The scented leather glove industry quickly became so important to the French that King Louis IX christened a guild of master glover-perfumers in 1651. Eventually, high taxes, diminished demand, and the dissolution of all guilds in France in 1791 spelled the end of the tanning industry in Grasse. Yet as demand for perfumes grew, the town focused on production of perfumes' core elements. From the ashes of the glove-making industry, the perfume industry rose — all thanks to tannins and Catherine de' Medici's gloves.

As noted earlier, tannins are versatile chemicals that first evolved in green algae as sunscreen and then evolved in plants as protective shields, toxic and otherwise. These molecules were transformed in nature and built a bridge from algal life underwater to plant life on land. We humans then put tannins to use to communicate our most important ideas across thousands of years.

Although we no longer use tannins to make leather or ink, we do use another chemical in this category as an everyday wonder drug. To understand why we use acetyl salicylic acid, or aspirin, as a medicine, we need to understand how the biosynthesis of its precursor chemical salicylic acid, evolved. The story begins, strangely enough, in the Andes, far away from where salicylic acid was pulled into the modern pharmacopoeia in Britain.

From the Andes to Aspirin

Like many tannins, aspirin is another phenolic chemical encountered by almost everyone. This over-the-counter drug is one of the best for reducing fevers (formerly called agues) and pain. Each year, we collectively swallow 120 billion bitter doses to treat these conditions and to prevent heart attack, stroke, and even colon cancer. That dosage is equivalent to 15 tablets a year for every human on the planet.

It wasn't modern medicine that gifted us with the powers of aspirin. We can still demark the moment aspirin's natural precursors entered the modern pharmacopoeia. In 1763, when the Reverend Edward Stone reported a connection between the "Peruvian bark," which had become the preferred antimalarial treatment in Europe, and the bark of a local willow tree:

> My Lord... There is a bark of an English tree, which I have found by experience to be a powerful astringent, and very efficacious in curing aguish and intermitting disorders. About six years ago, I accidentally tasted it, and was surprised at its extraordinary bitterness; which immediately raised me a suspicion of its having the properties of the Peruvian bark... The tree, from which this bark is taken, is... the common white willow.

Stone's discovery of the antifever properties of willow bark was inspired by the use of the so-called Peruvian bark from the fever tree to treat malaria and the intermittent fevers it causes. The fever tree is *Cinchona officinalis,* which is native to South America and is the principal source of the antimalarial drug quinine. Although I am temporarily diverting from the phenolics and flavonoids focus of this chapter by addressing the origins and uses of the alkaloid quinine, we will need this information to unravel the story of aspirin.

We now know that the twenty-three species of cinchona trees in South America produce quinine and related alkaloids. Proof that these chemicals deter and harm insects that ingest them comes from the laboratory.

In one experiment, researchers offered fly larvae a choice between a diet with or without quinine and followed their movements. The larvae all congregated away from the food with quinine. In a second trial, the larvae weren't given a choice. Those reared on the food with quinine weighed 33 percent less than did those reared on the control food.

Even more remarkably, when reared for five days on food containing natural concentrations of quinine, caterpillars of the African cotton moth weighed 90 percent less than did caterpillars reared on control food with no quinine. And of those fed the quinine-containing food, only 30 percent formed cocoons.

Of course, quinine is also toxic to the *Plasmodium* parasites that cause malaria, which is why it works to treat the disease. European colonists introduced malaria into the Americas in 1492, but the consensus is that the Incans had already discovered the fever-reducing properties of cinchona bark before the bark was brought to Europe in around 1633. The concentration of quinine in tonic water, used to make gin and tonics, is just not high enough to serve as a malaria preventative or treatment. We will return to quinine and its role in shaping geopolitics later.

You can now see why Stone was well aware of the healing powers of the cinchona. It had been used in Europe for more than a hundred years. This knowledge directly inspired him to experiment with the bark of the white willow, whose leaves and bark produced bitter phenolics known as salicylates.

It took ninety more years after Stone's remarkable but rudimentary clinical trial in 1763 until salicin, which breaks down into salicylic acid, was tweaked into a more active form in the body called acetylsalicylic acid. Bayer began selling it in 1899 under the name Aspirin.

This everyday wonder drug, the one medicine all of us have probably taken, owes its existence to both folk knowledge and the advances of modern medicine. Ultimately, there would be no aspirin without the ancient struggle that led plants like the willow to accumulate its precursor chemicals, salicylates, at high enough levels for them to be useful as protective shields.

Plants don't make salicylates for our sake. Yet evolution certainly did produce a pharmacopoeia that we have exploited over and over. So while nature might not have given us Eden, it did give us a wild Poison Garden.

The high concentrations of salicylates in willows evolved to protect them from stressors like herbivores and pathogenic microbes, long before humans inhabited the earth. Although all plants make salicylates, most don't produce them at levels high enough to be protective. Like the green tea catechins, it is the dose that makes the poison.

The most ancient role of salicylates in plants is the hormonal regulation of many processes, including growth and reproduction and defense against assault. One important function of salicylates is to activate an immune response in plants. Even the most ancient lineages of land plants, the mosses, hornworts, and liverworts, use salicin and its derivatives in this way.

The first time I sprayed experimental plants with a solution of salicylic acid, the breakdown product of salicin, I was an NIH postdoctoral research fellow at Harvard University. My mentors were evolutionary biologist Naomi Pierce and molecular biologist Fred Ausubel. The very next day, I tried to infect these plants with a pathogenic bacterium by injecting a suspension through openings in the leaves called stomata. But my plants were already resistant, thanks to the salicylic acid pretreatment they'd been given the day before. As a result, the bacteria couldn't colonize the leaves, which were primed to resist the microbial invaders. This immune-priming feature of salicylic acid was well understood in plants by then, but seeing it with my own eyes was richly satisfying.

However, it was not the salicylic acid itself that protected the plants. This chemical instead triggers a chain reaction that ends with the closure of the stomata to prevent bacteria from swimming into the inner sanctum of the leaf, where the sugars and proteins the bacteria need to grow are held. If that wasn't enough, the burst of salicylic acid signals to the plant's cells to produce potent antimicrobial toxins. After treatment with salicylic acid, many of the plant's tissues, even the parts that were not initially damaged, were protected from attack.

The evidence suggests that only much later in plant evolution did salicylates turn into actual toxins, a role they now play in some trees like the willow. This development leads to two important observations. First, salicylates the toxins evolved from salicylates the hormones—just as hydrolyzable tannins the toxins in plants evolved from hydrolyzable tannins the UVB-blocking algal sunscreens. In addition, the evolution of salicylates as toxins reminds us that it is the dose that makes the poison.

I was completely shocked to learn that, according to research reported in 2008, you and I and most other mammals have small amounts of salicylic acid regularly circulating in our blood. While most of the salicylic acid that can be detected in our blood comes from our diet, salicylic acid is found in the blood of people who hadn't consumed food containing it. Furthermore, when the salicylic acid precursor benzoic acid was given to test subjects, somehow, the body or associated microbes in the gut can make salicylic acid from this precursor. Sure enough, the test subjects given benzoic acid had detectable blood levels of salicylic acid that could not have come from their diets.

The scientists went a step further and found that the source was probably not the gut microbes. They detected salicylic acid in the blood of lab rats from a microbe-free lab rat colony. These rats were born via sterile cesarean section and kept on a germ-free diet that contained no salicylic acid. So, mammals seem to be able to make their own salicylic acid. The question is, why?

Salicylic acid, aspirin, and other nonsteroidal anti-inflammatory drugs like ibuprofen and naproxen (and possibly the chamazulene from chamomile) suppress the production of hormones called prostaglandins. Prostaglandins play a key role in turning on the inflammatory response in our bodies, causing inflammation and pain.

Inflammation can be a good thing. An adaptive response to injury or infection is essential to healing and recovery. However, there can be too much of a good thing. Chronic inflammation is a killer, leading to cancer, cardiovascular disease, dementia, diabetes, and an array of other serious health conditions.

Plants also make prostaglandin-like chemicals called jasmonates—named for the jasmine oil from which jasmonate was first isolated. Prostaglandins (in animals) and jasmonates (in plants) have nearly the same chemical structure. Jasmonates are produced by plants in response to attack by many herbivores and to physical damage, just as prostaglandins are produced by us humans and other animals in response to infection or injury. Like the salicylates that plants produce when they are assaulted by many microbes, jasmonates act as long-distance signals sent from the injured part of the plant to the uninjured parts, initiating the production of a toxin shield that can protect the whole plant from many herbivores.

The process recalls how Paul Revere instructed his compatriots to signal "one if by land, two if by sea" as the British began their invasion of the American colonies. Plants use both hormones to create the right response to the particular set of enemies invading at the moment. Salicylate production triggers the accumulation of antimicrobial toxins in the plant on the one hand, and jasmonate production leads to the accumulation of anti-herbivore toxins in the plant on the other. And to top it all off, salicylates suppress the production of jasmonates, and vice versa.

This biochemical interaction means that when plants are sprayed with salicylic acid, their jasmonate pathway is inhibited. The response is much like the way salicylates dampen prostaglandin production in us when we take aspirin, relieving our inflammation. Salicylate produced by plants also dampens the production of jasmonate, which is the plant's version of prostaglandin. This reaction can render plants susceptible to attack by herbivores, which can be resisted through the triggering of jasmonate-dependent defensive chemicals.

Both animals and plants can use salicylates in the same way—to dial down one branch of the immune system in favor of dialing up another. Why did plants and animals evolve to do this? We don't know for certain, but the organisms have limited resources and have evolved to economize. This efficiency leads to trade-offs—the notion of a jack-of-all-trades who is master of none applies to traits of other organisms in the natural world too.

Headaches in humans and bitter leaves of willows share something important in common. Plants and humans use many of the same or similar chemicals, like salicylates, jasmonates, and prostaglandins, to regulate their bodies because these living organisms share a common evolutionary ancestor. This commonality explains, at least in part, why so many chemicals produced by plants also have effects in us. The chemical messengers that the plant, animal, and human cells use to communicate with one another overlap. In many ways, a eukaryotic cell (namely, a cell with a nucleus and other discrete structures) is a eukaryotic cell.

On the other hand, plants and animals are wired in very different ways. Although both plants and animals use chemicals like dopamine to send chemical signals from one part of the body to another, from leaf to leaf or limb to limb, plants lack the nervous, musculoskeletal, cardiovascular, or digestive systems that animals have. Plants have also exploited these differences in wiring by producing some chemicals that may have no effect on the plant itself but have major effects on us. These chemicals include those that can taste bad, harm us, or even kill us. This other side of the evolutionary coin is just as important to examine if we are to understand the origin of nature's toxins and their intersection with our lives.

If there is one overarching lesson that I hope to impart from this book it is this: Evolution applies the similarities and differences between plants (and other organisms, like fungi and bacteria) and animals to various organisms' advantage. Plants have evolved to produce diverse chemicals that manipulate animal brains and brawn to their own ends, whether to keep us away or to draw us in. In turn, we have turned the tables ourselves and can tap into these chemicals as medicines and for other purposes.

Salicylic acid is a good example. All land plants seem to use salicylic acid as a hormone to regulate their responses to stressors. Humans and other animals appear to make small amounts, too, perhaps to regulate their inflammatory response. Salicylic acid is a poison to the animal enemies of plants like willow, meadowsweet, myrtle, wintergreen, and coca, all of which produce large amounts of its precursor, salicin. But for us humans, salicin and its derivatives like aspirin are among the most widely used medicines.

Salicylic acid is so pervasive that we can use it to follow a food-as-medicine approach, a concept I mentioned earlier. According to this idea, we sometimes use nature's toxins as medicines. Just by eating a lot of plant-based foods, for example, we can attain levels of salicylic acid on par with people following a daily low-dose aspirin regimen. So, you may have experienced the anti-inflammatory effects of this chemical without ever knowing it.

Salicylic acid levels are three times higher in rural vegetarian Indians than in Western vegetarians living in the United Kingdom. This difference can be linked to the higher use of spices, which are rich in salicylic acid, in the cuisines of rural India.

There may be potential health implications of this difference in salicylic acid levels. For example, the incidence of colon cancer is unusually low in these same vegetarian communities in rural Chennai, India, suggesting a link, however tenuous, between higher levels of dietary salicylic acid and lower colon cancer risk.

Much of the salicylic acid in the diet of these communities is delivered in the form of spices, some of which contained up to 1.5 percent of their weight in salicylic acid. The high content of this acid in spices does not necessarily explain why we use them, of course, but it is food for thought.

Is salicylic acid really the spice of life? Maybe not. There is also a downside to our use of aspirin. Ironically, some people take low-dose (baby) aspirin daily to reduce the risk of clots that could lead to stroke or heart attack. This practice is no longer recommended unless ordered by a doctor, because the potential for bleeding typically outweighs the potential for harm reduction. The same now goes for the use of baby aspirin to reduce the risk of colon cancer. In the United Kingdom alone, daily aspirin use kills roughly three thousand people a year because of the fatal gastrointestinal bleeds it can cause. Our use of nature's toxins requires that we walk a knife's edge between healing and harm. Again, we are reminded that these chemicals evolved not for us but because plants and the other organisms that make these substances in quantity want to live, too.

Humans aren't the only species to borrow nature's toxins to keep enemies at bay, treat disease, or gain access to new resources. As the earlier example of tetrodotoxin from newts, blue-ringed octopuses, and puffer fish illuminated, many animals borrow these chemicals from other organisms for their own benefit.

Across the Northern Hemisphere, a beautiful group of leaf beetles with black bodies adorned with orange spots feed only on the leaves of willows, poplars, and alders—all members of the family Salicaceae. Many species in this plant family produce high levels of salicin, which the willow leaf beetle larvae use as a precursor to make the toxin salicylaldehyde. This chemical, released from a beetle gland when predatory wasps attack, increases the chances of the beetle's survival. But predators can't say they weren't warned: the beetles have evolved their black and orange coloring to warn of the dangerous chemicals they contain.

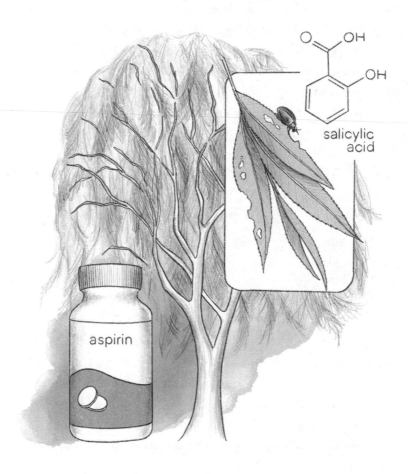

Another colorful and potentially poisonous animal is the banana slug. I am so enamored with them that Shane and I share our home with one named "The Colonel." I promise, there is a connection to aspirin, a potential one at least.

Schoolchildren in the San Francisco Bay Area and Pacific Northwest sometimes dare one another to lick these mollusks because something in the slug's mucus tingles the tongue and numbs the lips. The sensation has been described as Novocain-like. We don't yet know what this numbing substance is, but my colleagues and I are trying to solve the mystery.

I have a working hypothesis. A garden slug in Europe is known to secrete salicylic acid in the mucous trail on which it glides. A derivative of salicylic acid called methyl salicylate, or oil of wintergreen, activates our receptors for menthol flavors and causes sensations like those reported from the slug lickers. So, I am guessing that the investigative trail will lead to methyl salicylate as the numbing substances in banana slugs. If I'm wrong and we discover a new chemical that could be developed into a new anesthetic, all the better.

In addition to providing a mouth-numbing defense against predators, the slug's salicylic acid may manipulate plants into producing fewer anti-slug toxins by inhibiting the jasmonate pathway. Slimy!

The final class of chemicals I will cover in this chapter also holds the power to be a poison or a cure. You probably imbibed some of these flavonoids today, inadvertently. These are the furanocoumarins—the chemicals that make grapefruit juice a no-no if you are on certain medications.

Grapefruit and Giant Hogweed

The parsley family Apiaceae includes carrots, celery, cilantro, cumin, dill, fennel, parsley, and poison hemlock. The last-mentioned plant contains the alkaloid coniine, famous for allegedly killing Socrates. These plants, as well as those from the citrus, fig, and bean families, all make furanocoumarins, which are flavonoids.

Because furanocoumarins are contained in citrus, the FDA issued drug interaction warnings related to consumption of grapefruit juice. You may have heard about these warnings, or they may even affect you.

The activity of an important drug metabolism enzyme in our bodies, CYP3A4, is inhibited by furanocoumarins. The result is delayed drug deactivation if we ingest furanocoumarins. CYP3A4 is expressed in our small intestines and livers, and it helps us metabolize many toxins we ingest, whether they come from nature or the pharmacy. The delay in drug metabolism can be quite risky for those of us who take certain drugs regularly. If your body is slower than usual in eliminating a drug from your system, then when you take another dose the next day, the level of the drug in your body can become elevated, sometimes dangerously so. Furanocoumarins can slow the rate of detoxification of drugs by suppressing the activity of CYP3A4.

Although nobody thinks of grapefruit juice as toxic, it can enhance the toxicity of many prescription drugs and even some over-the-counter ones. A twenty-nine-year-old man from Michigan died suddenly from a heart arrhythmia brought on by toxic levels of the over-the-counter antihistamine terfenadine (sold as Seldane in the United States and Triludan in the United Kingdom) in his blood. The two glasses of grapefruit juice he'd consumed that morning suppressed his CYP3A4 activity and led to elevated levels of the usually harmless drug in his body, which in turn triggered the fatal heart attack. In the wake of his death, the drug was immediately removed from the market.

But other drugs whose metabolism in our bodies is influenced by the furanocoumarins in grapefruit juice are so essential that we cannot simply do away with them. Examples of these medications are cyclosporine antibiotics, opioids, and statins. In all, there are eighty-five over-the-counter and prescription drugs whose rate of elimination from our bodies is influenced by the furanocoumarins in grapefruit juice.

Furanocoumarins can also cause third-degree chemical burns through a completely unrelated process known as phytophotodermatitis. These burns occur when skin that has come into contact with furanocoumarins is exposed to UVA light from the sun. The furanocoumarins activated by this light make their way into the skin cells and bind to DNA. The modified DNA causes skin cells to die and drives an inflammatory response of blistering, peeling, and discoloration. When furanocoumarin exposure is chronic, skin cancer risk is elevated as well.

Although I don't want to scare you, a thirty-year-old man was badly burned after squeezing the juice of sixty limes for margaritas and then exposing himself to the sun. He recovered, but this skin disease is a serious occupational hazard for citrus, celery, and parsley farmers as well as for grocers and bartenders too! Another case involved twelve children who were burned after being exposed to the sun while making lime pomanders for the holidays.

The giant hogweed (*Heracleum mantegazzianum*) is perhaps a bigger threat to public health. This herbaceous plant is native to Russia and the Republic of Georgia and has become a toxic, invasive species elsewhere. Cultivated as an ornamental, it easily escapes into the wild, where unintended encounters with its sap, which contains high concentrations of furanocoumarins, cause countless cases of phytophotodermatitis. Burns have even led to the need for skin grafts and limb amputations.

Like many of the other toxins covered in this book, the poison can also be the cure in the right context. Knowledge of how a substance can pivot from bad to good almost always flows from Indigenous and local knowledge holders and practices. So it is with furanocoumarins. Psoralen, the principal furanocoumarin that is found in plants like giant hogweed and can cause so much harm, is also a highly effective treatment for skin disorders that can cause social stigma. In fact, furanocoumarins have been used in so-called photomedicine for thousands of years—first emerging long before the word *phytophotodermatitis* was coined in 1942.

In the Mediterranean region, a treatment for vitiligo, the loss of skin pigmentation, first appeared in the Papyrus Ebers 3,500 years ago. The treatment involved the greater ammi plant *Ammi majus* (false Queen Anne's lace), which still grows along the Nile. This same plant is described in the thirteenth-century writings of Ibn al-Baytar, a physician from Al-Andalus, the Muslim-ruled area of Spain. Building on knowledge from Egyptian herbalists who still used greater ammi in their practices, University of Cairo pharmacologist I. R. Fahmy and collaborators first isolated psoralen, the active chemical from the greater ammi, in 1947 and then conducted the earliest clinical trials using it to treat skin disorders. Psoralen taken orally or topically plus UVA treatment treats eczema and psoriasis by slowing the overgrowth of skin cells, allowing skin lesions to resolve. It is also used against cutaneous T-cell carcinoma. A clinical trial is under way in which X-rays rather than UVA light are used to activate psoralen that's been delivered deep into solid cancer tumors, regions not accessible by UVA light. Thanks to these efforts by Egyptian scientists, furanocoumarins entered modern medical usage and continue to be a highly efficacious treatment for certain skin disorders.

The ancient use of extracts of greater ammi as a skin-darkening agent to treat vitiligo inspired its application as a tanning activator in Europe. Psoralen was used sometimes in conjunction with tanning beds, but this practice was banned in the mid-1990s because it increased skin cancer risk.

Just as distantly related plants like some members of the parsley family Apiaceae and some citrus in the family Rutaceae produce furanocoumarins, diverse Indigenous cultures also tap into the same chemicals over and over. The dried fruits of *Psoralea corylifolia,* a bean that also produces psoralen, independently appear as a treatment for vitiligo in ancient Ayurvedic and Chinese medical texts written between the twelfth and tenth centuries BCE. Cow parsnip, which also produces psoralen, was used as a poultice for boils and other skin ailments by many Indigenous tribes and First Nations of North America.

The ultimate question is, why do plants bother to make furanocoumarins in the first place? As you have probably guessed, the evidence shows that furanocoumarins evolved in plants as highly effective antiherbivore defenses. Furanocoumarins are potent insecticides through the same actions that cause skin problems—the chemicals are cell-destroying toxins.

Yet nature does find a way around furanocoumarins. For example, entomologist May Berenbaum discovered that caterpillars of both parsnip webworms and swallowtail butterflies have independently evolved ways to pierce the chemical shield of furanocoumarin-containing plants long before humans existed. The trade-off is that these caterpillars can only eat the plants that make furanocoumarins.

Just like us, moths and butterflies use the otherwise-toxic furanocoumarins to their benefit, albeit indirectly. By being able to overcome these toxins, the insects are allowed access to the nutrients that are otherwise protected by the furanocoumarins in the plants. These specialized winged creatures may face less competition from other animals for food because this food is toxic to most other herbivore species. Berenbaum has used this ongoing dynamic between furanocoumarin-bearing plants and the insects that attack them as a case study in coevolution between species.

So, furanocoumarins can be both a cause of, and a treatment for, skin disorders. But they first evolved as toxins for the plant to wield a defense against attacking herbivores, and this shield can be pierced by some specialized insects that evolved ways of overcoming the toxins.

Phenolic and flavonoid chemicals like tannins, salicylic acid, and furanocoumarins represent an enormous and important class of nature's toxins. However, terpenoids, such as tetrodotoxin and saponins, include an even larger set of toxins. We'll look at these toxins in more depth in the next chapter.

3.

Toxic, Titillating, Tumor-Killing Terpenoids

All the plants of a given country,
all those of a given place,
are at war with one another.

—Augustin Pyramus de Candolle, "Géographie
botanique," *Dictionnaire des sciences naturelles*

Bays, Blues, and Ball Games

My father died from complications of alcohol use disorder (AUD) in the winter of 2017. In the days after his death, I found myself in the wintry quiet, shinrin-yoku style under the boughs of ancient moss-covered, mushroom-studded coast redwood and California bay trees in the hills above our home in Oakland.

One such moment in the forest led to the first glimmers of this book. It is when I really began to understand how the similarities and differences among the mosses, mushrooms, trees, and me are a dialectic that could be used to better understand not only my science but also my father's death.

Squinting skyward on one of those days, I tried to make out the crowns of the trees, but they were shrouded in fog. Cold drops landed on my face from fog-soaked branches above, connecting me with the frigid San Francisco Bay below. I crushed some coast redwood needles and California bay laurel leaves in my hand and inhaled deeply. I got what I had come for—a dose of what

Robert Louis Stevenson called "frosty pepper up my nose" from the cocktail of terpenoids contained in the leaves of each species. Tears instantly welled, reminding me that I was still alive despite the numbness I felt.

Terpenoids are the largest chemical family made by life on earth. More than eighty thousand compounds have been identified in nature. Like phenolics and flavonoids, terpenoids include some of the most ancient biological molecules, including the sterols preserved for 538 million years in *Dickinsonia* fossils. The now-extinct *Dickinsonia*, the oldest animals that have been found in the geologic record, evolved before the first land plants.

Terpenoids are essential for life as we know it. The cholesterol in our cell membranes is a terpenoid, as is the vitamin A that captures light in our retinas. So important are terpenoids that they played a role in the origin of the first cells that evolved four billion years ago and are the oldest biological molecules known from fossils.

Terpenoids synthesized by plants are used by all of us when we clean, drink, eat, take medicine, recreate, relax, and practice spirituality. From alpha-pinene to atropine, camphor to cannabinoids, menthol to myristicin, and Taxol to thymol, endless varieties of terpenoids spice up our days, calm our nights, and even extend our lives. Of course, there is a dark side: they can also harm or even kill us.

A five-carbon molecule called isoprene forms the backbone of all terpenoids. Isoprene is exhaled in every breath, and a hundred billion kilograms, an atmosphere-influencing quantity, is released by land plants each year globally. Our bodies make isoprene as a by-product of cholesterol biosynthesis, and the amount emitted can even be used to determine a person's cholesterol levels.

We emit so much isoprene that spikes are detected in the air above soccer stadiums when the crowds cheer, and the level rises in cinemas during climactic scenes. Although plants produce isoprene during metabolism just as humans do, this chemical is transformed from trash into treasure in plants when temperatures soar.

Isoprene protects leaves from excess heat generated by the sun's rays. When plants release isoprene into the air, the molecule reacts with oxygen radicals to form aerosols that produce a distinctive blue color in the lower atmosphere. The Blue Ridge Mountains of the United States, the Blue Mountains of Jamaica and Australia, and the Cordillera Paine (*paine* is Tehuelche for "blue") in Chile may be so named, thanks to the isoprene-pumping forests blanketing their flanks.

Leonardo da Vinci was so enthralled by the blue air ensconcing the Tuscan hills, he used them as the backdrop for his *Portrait of Lisa Gherardini, wife of Francesco del Giocondo,* or, as most of us know it, the *Mona Lisa.*

Isoprene doesn't always paint such a pretty picture. It reacts with nitrogen oxides from combustion engines, forming ozone, which acts as a greenhouse gas and contributes to smog, which stings the eyes and burns the throat. The amount of isoprene released from plants each year is equivalent to the annual amount of methane put into the global atmosphere from all sources, natural and human. The line between purple mountain majesties and purple haze is razor thin.

Natural rubber is also made of isoprene molecules linked together to form a polymer called *cis*-1,4-polyisoprene. It is now produced almost exclusively from the milky latex of the Pará rubber tree, a South American member of the spurge or poinsettia family. Hours after the rubber tree's liquid latex is exposed to the air, it polymerizes and turns into solid rubber, which is vulcanized for commercial use through the application of heat and sulfur. Travel via airplanes, cars, motorcycles, trucks, and bicycles wouldn't be possible without natural rubber, which constitutes up to 50 percent of most modern tires. Indigenous peoples of South America were the first to use Pará rubber from wild trees growing in the rain forest.

Separately, the Panama rubber tree, a species in the fig family, was tapped to produce the heavy rubber balls of a Mesoamerican ball game, an ancient sport that played a central role in these societies. The oldest stone ball court known today was a grand one, built a staggering thirty-four hundred years ago at the Paso de la Amada in Chiapas, Mexico.

It was through this Mesoamerican ball game that natural rubber was first brought to the attention of Europeans in 1510. Ultimately, the development of this natural rubber product spurred the mass production of baseballs, basketballs, footballs, golf balls, hockey pucks, rugby balls, soccer balls, and tennis balls in the nineteenth and twentieth centuries.

Why do rubber trees and vines go to all the trouble of making rubber-laden latex under their bark in the first place? We certainly didn't breed them to do so. Latex serves no known function in plants other than as a defense.

More than twenty thousand plant species and, through convergent evolution, milk-cap mushrooms produce latex. In these plants and mushrooms, latex acts against herbivores and microbes by trapping them and then sealing off the wounds they created. Latex itself therefore serves as a defense, but it is also a toxic conduit for the delivery of even more potent natural toxins, including terpenoids.

The Congo rubber vine, in the dogbane family, also produces rubber. In parts of Central Africa, the rubber is traditionally used to adhere poisons to arrowheads and to treat decaying teeth, among many other uses.

Millions of people in what is now the Democratic Republic of the Congo were harmed and many killed in pursuit of rubber to meet the demands of colonial Europe, Canada, and the United States. Whether the product came from South American Pará rubber trees cultivated on African soil in the twentieth century or wild Congo rubber vines tapped by King Leopold II of Belgium, the human suffering caused by rubber production is a permanent blemish on all of Western society. For example, at Leopold's behest, when enslaved Congolese laborers collected latex from wild Congo rubber vines, the latex that oozed directly onto their skin would harden, and had to be painfully scraped off, hair and all.

Although native to the Amazon rain forest, the Pará rubber tree is now mostly cultivated in Southeast Asia. An early act of apparent biopiracy by the British Empire, coupled with a leaf blight in the plant's native range in South America, may have set this agricultural transplantation in motion. In 1876, Sir Henry Alexander Wickham is said to have smuggled seeds out of Brazil to botanists at the Royal Botanic Gardens, Kew, just west of London. Some of the rubber tree seedlings that germinated were planted in British colonies in Southeast Asia, including Malaya and Singapore, creating a British monopoly on natural rubber.

This juggernaut and the blight ruined the rubber-dependent economy of Brazil, which was built on the backs of heavily exploited workers. It also led to a major problem for the Allies when Japan invaded Malaya the day before the Japanese attacked Pearl Harbor because the supply of rubber was cut off.

The United States could not have won World War II without access to new sources of rubber. My own life intersected with the US pursuit of alternative sources of natural rubber. As an eight-year-old, I visited Seminole Lodge, the estate of Thomas Edison in Fort Myers, Florida, with the rest of my family. It was the first time I'd been exposed to the life of a scientist. My only memory of the trip is of the massive fig tree on the property that read: "Banyan Tree — Given to Edison by Firestone in 1925 — Circumference of Aerial Roots 390 ft."

I later learned that it was Harvey Firestone, the tire magnate, who had brought the tree from India and given it to Edison in the hopes that its milky latex would soon provide a domestic source of natural rubber on the US mainland. Edison analyzed more than seventeen thousand latex-producing species for their potential as sources of natural rubber. The research was funded in part by Firestone and Henry Ford, both of whom were troubled by the rubber monopolies held by the British and Dutch in the early twentieth century. The inventor found that more than a thousand of these species produced measurable rubber in their latex.

Edison died six years after receiving the tree. Around the same time, synthetic rubber from petroleum was invented to augment the supply. Still, natural rubber remained a $1.6 billion industry in 2021, and nearly all of it is produced in Southeast Asia.

The tour of Edison's research center, which was devoted to understanding the nature of chemical defenses of plants, was providential in hindsight. (Not surprisingly, the now-documented antisemitic views held by Ford and Firestone, and the facts surrounding the dystopian Brazilian rubber town of Fordlândia, were not mentioned.) I would go on to study the cardiac glycosides in the latex of milkweeds thirty years later.

As I learned firsthand when growing plants that produce latex, the substance is stored in long cells called laticifers under high pressure in anticipation of attack: a goo-filled gun, locked and loaded by evolution. This loaded-gun arrangement can lead to unpleasant surprises.

Early last year, shortly after two of my friends moved to New Mexico from San Francisco, I received an urgent text message from one of the pair asking: "Should we be worried if he got a drop of cactus juice/milk in his eye?"

It was a familiar motif. I have come to accept that my friends and family show me their affection by tapping into my arcane knowledge of nature. After reminding them that I'm more like Doctor Doolittle and cannot give any medical advice, because I am not a physician, I can usually at least identify the offender and maybe even the toxins at play.

Thinking it was a cactus, I replied no. But then she texted that he had flushed his eyes with water because they had begun to burn. I asked for photos of the plant because I began to doubt the accuracy of their identification. He had been moving the potted plant as they set up their new place, she said, but the top of the stem accidentally hit the ceiling and snapped off, and, as it did, "liquid" from the broken stem shot forth and landed in his eyes.

We had been fooled by convergent evolution—the independent origin of the same trait. After one look at the actual photo, I saw that the plant was actually a spurge in the genus *Euphorbia*, native to Africa. To the untrained eye, the plant looks almost identical to a cactus, all of which are native to the Americas. *Euphorbia* is a diverse genus that includes houseplants and garden plants in the same family to which the Pará rubber tree belongs.

My botanist friends and I like to call plants like these members of the "Houseplantaceae," an eclectic group of mostly rain forest understory and desert plants often found in the living rooms, offices, hotel lobbies, shops, dorms, and kitchens of our lives. Though they may seem mundane, as undomesticated plants plucked from the war of nature, their tissues are often toxic to varying degrees, whether the toxicity applies to humans, pets, or both.

I was concerned. I knew that my friend's eye might require immediate medical treatment. The terpenoids in the latex from *Euphorbia* can chemically burn the cornea and, left untreated, can even lead to blindness through secondary bacterial infection.

I quickly texted back: "I think it might be a spurge. It's not a cactus. They do have a sap that is very bad, I think. You may want to proceed to taking him to urgent care. I wouldn't hesitate. Make sure you tell them that this is not a cactus."

He was rushed to urgent care, his eye was saved from permanent damage, and he has completely recovered. Ever the professor, I couldn't help but use this moment as a teachable one: "Not to make light of this at all," I texted, "but plants don't wanna be attacked. The best offense is a good defense." She responded a few hours later: "I learned something new about the universe today."

We'll now consider the terpenoids in the amber of my ring and the terpenoid that hastened the death of my father. Each of these toxins was forged in nature too.

Balsams, Birches, and Beers

Many terpenoids serve as the basic building blocks of life, but as the latex-oozing rubber trees and toxin-squirting spurges demonstrate, terpenoids also carry a message. Alpha-pinene, which I sought out the day I was forest bathing under the coast redwood and California bay trees, is one of these terpenoids. Carried in the air as an aromatic volatile, it made its way into my nose and brought a smile to my face. Human delight is not why this molecule evolved.

The redwood's redolence and flat needles that crackled underfoot transported me thirty-five years back in time to a trail lined with feathery balsam firs. That day, my father, brother, and I walked in the boreal forest around our house in northeastern Minnesota at the edge of the Sax-Zim Bog. Unlike the redwood's thick, shaggy bark, the fir's was thin and smooth, except for the blisters every few inches.

My father pierced a few blisters with a stick. A thick, clear resin burst forth and oozed down the trunk. More than a parlor trick, the resin could be used like lighter fluid to start an emergency fire. It was important that we learned survival skills up there, where winter temperatures dropped to minus forty degrees, the point at which Celsius and Fahrenheit scales intersect.

Tree resins evolved long before latex did. They are composed of fatty acids, phenolics, and monoterpenoids (two isoprene units), sesquiterpenoids (three isoprene units), and diterpenoids (four isoprene units), and a much simpler chemical composition than that of latex, although the function of resins — to dissuade, maim, and kill — is similar.

Monoterpenoids and sesquiterpenoids are volatile and facilitate the fluidity of resin as it flows, whereas diterpenoids create the hardened resin. The resin and essential oil of balsam firs, coast redwoods, California bays, and many other plants contain alpha-pinene, a monoterpenoid used to make air fresheners, candles, floor polish, and turpentine. Turpentine takes its name from the Greek *terebinthos*, meaning "tree resin."

Why do balsam fir trees produce alpha-pinene-infused resin? It certainly isn't to delight us with its smell, to provide emergency heat, or to make the strings of a Stradivarius sing. Resin, like latex, evolved in plants hundreds of millions of years before the first ancient humans walked on two legs. Balsam firs use resin to entrap and poison attackers like insects and microbes.

As a kid, I occasionally found insects trapped in candied resin drops on fir trunks. It called to mind the *Jurassic Park* mosquito that was sealed in an amber tomb, along with the apocryphal blood meal taken from a dinosaur.

Amber is simply ancient tree resin that hardened with exposure to air and was then buried in rock. The oldest amber known has been preserved for 320 million years.

There is even more ancient evidence of terpenoids being used as chemical defenses in plants in the form of fossilized oil body cells. These cells are found in liverworts, plants that are close relatives of mosses. The oil body cells contain terpenoids that help protect liverworts from attack by insects. Liverwort-like plants were also the first plants to have evolved from green algal ancestors. Paleontologists Conrad Labandeira, Susan Tremblay, and their collaborators recently discovered that fossilized liverwort plants from 385-million-year-old rock indeed had oil body cells. This evidence of oil body cells in liverworts that lived in the Devonian period suggests that defensive terpenoids have been made by plants since the beginning. Oil bodies were an important defense innovation in land plants because invertebrate animals were on land by then too, and animals with backbones were not far behind. The "fishapod" *Tiktaalik* was already lurking in the shallows, presaging our four-limbed ancestors' arrival on land.

Given how well many terpenoids protect plants from attack, it is hard to imagine anything eating alpha-pinene-infused needles, but some caterpillars can. Although most of us associate caterpillars with the larvae of butterflies and moths, caterpillars are also found in the sawfly family, which falls into the same lineage as ants, bees, and wasps.

Some sawflies are major crop pests, so named because of the eponymous organ used by females to place eggs in plant tissue. Pine sawfly caterpillars only eat needles and even store alpha-pinene and other turpentine derivatives in special pouches near their heads. If a predator tries to attack, deadly droplets are drawn up to the mouth and strategically lobbed onto the attacker's body, in the same way that hot tar pitch was dropped onto enemies from the machicolations in medieval castles.

Pine sawflies have turned the original function of the resin—a plant defense—on its head to use it for their own devices, just as we humans do. Mountain pine beetles take it a step further. These pests are ravaging North American pine forests amid a worsening drought in the rapidly warming climate.

To gain the upper hand, the beetles use alpha-pinene and ethanol vapors emitted by naturally weakened trees as host-finding cues. It seems odd that plants would make ethanol. But they do so in response to oxygen deprivation and water stress, just as we produce lactic acid (which causes that burning feeling in our muscles) during exercise.

Once a mountain pine beetle arrives at a host tree, the insect begins to attract other mountain pine beetles using an aggregation pheromone called *trans*-verbenol. This pheromone is actually made from the alpha-pinene in the wood the beetles consumed as larvae in their natal tree. The piney perfume is a clarion call that triggers a mass attack of mountain pine beetles on a single tree.

This aggregation of beetle bodies might seem counterintuitive. After all, why would a mass attack give an individual beetle an advantage instead of creating competition? When they bore into the trunk, mountain pine beetles release spores of blue stain fungi carried from their natal trees. The fungi that evolved resistance to the resin grow in the wood and block the weakened tree's resin canals or ducts, neutralizing one of its primary defenses. The more, the merrier.

In Colorado, where I conduct research in the summer at the Rocky Mountain Biological Laboratory on plant-animal interactions, it is common to find furniture and wood paneling made from the wood of infected trees. Such "blue stain" pine wood is a novelty that can be traced back to alpha-pinene, the most delicious poison produced by trees as a defense and then co-opted by beetles to attract their kind to colonize and kill their host trees.

But there can be too much of a good thing for the mountain pine beetles. Though they use alpha-pinene as a chemical lure, if a tree produces high-enough levels, the chemical can thwart the beetles' attempts at colonizing. In fact, the oldest living trees in the world, bristlecone pines, have alpha-pinene to thank for their continued persistence, at least for now.

Bristlecone pines, including the 4,853-year-old Methuselah Tree living in my home state of California, make up to eight times more alpha-pinene than do other pine species in the southern Sierra Nevada and Great Basin. The extra alpha-pinene may explain why fewer bristlecone pines in the region have died in the past decades from mountain pine beetle infestations than other pine species that grow alongside them and have been ravaged.

From tree ring analyses, which document the lignified rings that mark the yearly growth you have seen in cross sections of trunks, we know that mountain pine beetles have been attacking pines for millennia. But the beetles have not damaged bristlecone pines so much; these trees seem to have gained the upper hand—at least for now—in part, it seems, from their high production of alpha-pinene laden resin. However, with warming and drying conditions, the tide may turn and favor the mountain pine beetles.

Alpha-pinene is both an attractant and a deterrent to humans and other animals, depending on the dose and the target. As a deterrent, alpha-pinene can be deadly, even to us. For one tragic example, as tractor-trailer driver Ricardo Garcia was cleaning out the inside of his tanker after delivering a load, he suddenly collapsed. Although he was quickly pulled out of the tanker by coworkers, he wasn't wearing a respirator. Sadly, Garcia died at the hospital from excessive exposure to the same vapors I sought under the boughs of the coast redwoods and California bays to calm my mind.

Another tree species near and dear to my heart also uses terpenoids to its advantage. Intermingled with the balsam firs along the trail we used to walk in Minnesota were paper birches. My father showed my brother and me how they too could be used to build a fire. Even if the outside surface of the birch bark was damp, its dry underlayers ignited readily, owing to the waterproofing terpenoids in the outer layer of bark.

Birch trees hold a special place in the cultures of Indigenous peoples of the high latitudes of North America, Europe, and Asia, often serving as trees of life, axes mundi that connect heaven and earth. The poem "Birches" by Robert Frost captures this essence:

> *I'd like to go by climbing a birch tree,*
> *And climb black branches up a snow-white trunk*
> *Toward heaven, till the tree could bear no more,*
> *But dipped its top and set me down again.*

Their alabaster trunks, waterproofing, and flammability are derived from a terpenoid known as betulin, which can compose up to 35 percent of the bark extract. Known since 1788, betulin is one of the first plant chemicals to have been isolated. Isolating this chemical probably wasn't so difficult—if you have ever rubbed a birch trunk with your hand or picked up a birch log to throw on the fire, the white powdery substance left on your palms was betulin. More recently, betulin derivatives have shown promise as anticancer, anti-inflammatory, and antiviral medicines.

Betulin's light-reflecting properties do more than whiten. They protect the thin layer of living cells inside the trunk from the sun's withering rays at high latitudes in winter. Keeping cold in winter might seem counterintuitive, but doing so prevents dangerous freeze-thaw cycles that would rupture cells in the trunks, just as such cycles burst pipes. Betulin also deters attack by microbes and herbivores.

Of course, we don't know if these protective qualities are the ultimate cause of betulin's evolution. But these are some of its biological functions now, and bark containing a high concentration of betulin has clearly been adaptive for birch trees in one way or another for a long time.

Modern scientists were not the first to discover the usefulness of balsam fir and paper birch, of course. In North America, the Anishinaabe—original people—of northeastern Minnesota rely on both these trees and have done so for generations. Paper birch bark lines the roofs of wigwams, sugar boxes to collect maple syrup, and canoe hulls, and balsam fir pitch is applied to seal the bark.

The rigid yet smooth white bark of the paper birch is also waterproof and rot resistant. All these qualities facilitated its use as a medium for pictographs across the generations as well. The ancient and sacred *wiigwaasabakoon*, birch bark scrolls, may represent the oldest pictographic language in North America. *Wiigwaasabakoon* record historical events, legends, maps, and rituals, and serve as *mide-wiigwaas*, communicating the practices of the Midewiwin priesthood, or Grand Medicine Society. Hidden in caves or hollow trees or buried underground, *wiigwaasabakoon* resist decay, thanks in part to betulin.

Once we moved an hour north of Duluth and away from the river, my father seemed equally calmed during our weekend walks on the trails through the balsam fir and paper birch forest. There, he brought us into the fold of his knowledge of nature.

But after sunset, his nightly twelve-pack of beer transformed him into an entirely different man, stunted and incoherent. Jekyll-and-Hyde dipoles seemed to flip with the terpenoids from the trees during the day and the hops-infused alcohol at night. Although he took it to the extreme, he was simply doing what we and our ancestors have done for the last fifty thousand years or more: using nature's toxins to transcend suffering.

The problem, as you now know, is that nature's pharmacopoeia was not invented for us. Its chemicals, like the ethanol we use to make alcoholic beverages, evolved long before the biosphere became aware of itself through human consciousness. There is no design with us in mind and no guarantee that the good will outweigh the bad.

A Toxic Larder

Ethanol is unique among nature's toxins because it cannot be easily placed into one of the chemical classes that are usually used to categorize the poisons we are discussing. Conveniently, though, many organisms convert ethanol to mevalonate, a precursor to terpenoids in the terpenoid pathway, so ethanol actually fits right in with terpenoids.

Despite my father's AUD, and as a light drinker myself, I cannot help but agree with Shakespeare's aphorism that "good company, good welcome, good wine, can make good people." *I* certainly think I am more fun when I've had a glass of champagne, but only one.

At the same time, although alcohol is the most widely used social lubricant, the consensus is now that overall, *any* alcohol consumption, even one drink a day, carries health risks, including higher risk of cancer, liver disease, heart disease, and death in accidents. However, for people over forty who are at higher risk of cardiovascular disease in certain populations, around one-half of a drink per day is associated with protecting against heart attacks. Still, Health Canada now recommends no more than two drinks per week, given that the costs outweigh the slight cardiovascular benefit.

AUD is the third-leading cause of preventable death in the United States, the first being tobacco use, and the second, poor diet and lack of exercise. Binge drinking kills more than forty thousand people per year in the United States alone.

Still, I plan on having one glass of champagne to celebrate New Year's Eve with Shane. I partake despite the fact that I know that AUD has found its way into every branch of my family tree.

Ethanol, like all the natural toxins we've discussed, didn't evolve with us in mind. Brewer's yeast, which humans domesticated from wild fruit-associated strains, efficiently ferments sugars into ethanol, hence the genus name *Saccharomyces*, or "sugar fungus." Yeast's ability to make alcohol evolved long before humans were around, likely as a means for these fungi to survive oxygen deprivation deep in rotting fruit. In the absence of oxygen, yeast can burn energy from sugar if they first convert the sugar into ethanol.

Brewer's yeast is resistant to the toxic effects of ethanol, while most other microbes are not. So one way of looking at this is that yeast can use the ethanol it makes as a defense against competitor microbes that colonize the fruit too. So, for brewer's yeast, ethanol is a poisonous private reserve of energy—a toxic larder. But there's a limit to their resistance: when ethanol levels exceed 20 percent, even brewer's yeast cells will perish in their own home brew.

The *Drosophila melanogaster* fruit flies I study live through the toxic niche carved out by brewer's yeast. As the old joke goes, "Time flies like an arrow; fruit flies like a banana." And not just any banana will do. The insects prefer ripe fruit that hosts ethanol-producing yeast. Unsurprisingly, fruit flies are resistant to low concentrations of ethanol, like brewer's yeast but unlike most other insects. In certain concentrations—around 3 percent alcohol by volume—consuming ethanol even extends their life span. Higher concentrations shorten their lives.

But the cost-to-benefit ratio flips if parasitoid wasps are lurking. Using a syringe-like structure emerging from its midsection, a female parasitoid wasp injects a single egg into the body of a fruit fly larva, along with a dose of venom and viruslike particles that suppress the fly's immune system. After hatching in the fly larva, the wasp larva consumes the host just after the fly larva forms a puparium, which is the fly version of a butterfly's chrysalis. The wasp uses this borrowed pupal case to metamorphose into an adult. Instead of an adult fly emerging from the puparium, an adult wasp emerges from the sarcophagus.

Using its immune system, the fly larva can sometimes kill the wasp's egg before it hatches. If that doesn't work, the wasp egg can be pickled in the blood of the fly by the ethanol the fly larva has consumed, but only if ethanol concentrations are at roughly the same as wine, or around 10 to 15 percent. Talk about stress drinking!

Because the fly larva must eat the food near where it hatched—it is a maggot, after all—its mother's ability to discern low from high ethanol concentrations is key to its survival if wasps are a threat. Indeed, fruit fly mothers prefer to lay eggs in fermenting fruit with higher ethanol concentrations, but there is a catch. They only do so if they have actually *seen* parasitoid wasps nearby. The cost of a high ethanol diet is a shorter life for the fruit flies. Nonetheless, in the presence of wasps, it's a cost worth bearing—as we've seen, the alternative is worse.

Evolution has threaded this needle by endowing flies with chemical receptors for ethanol and the ability to discern threat levels from enemies. The flies can weigh the threat level posed by toxic ethanol on the one hand and deadly parasitoids on the other.

Parasitoids are so named because they are infectious agents, but they *must* kill their hosts to complete their own development, just like the xenomorphs in the movie *Alien*. In contrast, parasites don't always kill their hosts to complete their life cycles. The mother parasitoid wasps use a complex brew of venom and other factors to suppress the host's immune system so that their offspring begin to eat it alive from the inside.

To Charles Darwin, parasitoids were proof positive of evolution. Darwin saw creatures on their own terms. In his view, the suffering of a fly larva or a caterpillar helped frame the larger argument of evolution.

We know Darwin held this view because on May 22, 1860, he wrote a letter to his confidant, Harvard botanist Asa Gray, a devout Christian. At first, the letter discusses reviews of his new book, *On the Origin of Species*. Eventually Darwin turns to an acerbic book review by a theologian and uses the existence of parasitoids to counter the theologian's argument. Darwin concludes that the existence of the Ichneumonidae, a family of parasitoid wasps that mostly attack caterpillars, was proof that evolution, not the hand of a creator, could produce such beasts:

> With respect to the theological view of the question; this is always painful to me. —I am bewildered. —I had no intention to write atheistically. But I own that I cannot see, as plainly as others do, & as I should wish to do, evidence of design & beneficence on all sides of us. There seems to me too much misery in the world. I cannot persuade myself that a beneficent & omnipotent God would have designedly created the Ichneumonidae with the express intention of their feeding within the living bodies of caterpillars, or that a cat should play with mice. Not believing this, I see no necessity in the belief that the eye was expressly designed. On the other hand I cannot anyhow be contented to view this wonderful universe & especially the nature of man, & to conclude that everything is the result of brute force. I am inclined to look at everything as resulting from designed laws, with the details, whether good or bad, left to the working out of what we may call chance. Not that this notion *at*

all satisfies me. I feel most deeply that the whole subject is too profound for the human intellect. A dog might as well speculate on the mind of Newton. — Let each man hope & believe what he can. —

The letter conveys that Darwin's struggle to understand the apparent cruelty inherent in the war of nature was part of the same struggle to understand the meaning of life itself. His last line was particularly poignant because he concedes what most of us know, deep down — we are looking through a glass, darkly. While we have made so much progress in discerning the truth about the physical universe, including how life evolved, mystery endures.

Just as the parasitoids helped Darwin frame an argument on evolution, there is an important lesson in the toxic larder made by brewer's yeast. Both the yeast and the fruit fly take advantage of a toxin in their environments for food and to keep competitors at bay. Too little or too much of the toxin, and they will lose out. To win the struggle for existence, each must balance the benefits and costs of using ethanol, which is to walk along a knife's edge. The question is, are we really that different? In some ways, it doesn't seem like it.

My father used ethanol as his nightly "medicine." Ethanol is thought to mimic gamma-aminobutyric acid (GABA), the amino-acid-derived neurotransmitter that the brains of all animals use to dampen the activity of the nervous system. It may actually bind to the $GABA_A$ receptors themselves. Because $GABA_A$ receptors dampen brain activity when they are activated, we often feel sleepy after a drink. Drinking relieved my father of serious nerve pain he suffered from a motorcycle accident before I was born.

Ironically, he had been hit by a drunk driver who ran a red light. The car plowed through the intersection and didn't stop after it hit him. To avoid being hit by other cars, my father grabbed hold of the vehicle and was dragged for blocks. His nose was nearly completely torn off, and the nerves in his neck permanently damaged.

His large daily dose of ethanol probably also helped ease the effects of the many other concussions he'd had, including one from another car accident and many from junior high and high school football games. In one of our last conversations on the phone, he explained through tears how, as a teenager and star football player at a prep school, he had awoken to find a Roman Catholic monk sexually abusing him in the recovery room after he had been knocked out in a football game. He began to weep and then abruptly hung up.

In addition to coping with physical trauma, drinking is a common coping mechanism following sexual abuse. Remarkably, he was quite frank about the fact that he'd consumed more than a hundred thousand beers in his life to ward off these and whatever other demons haunted him. He recounted that his doctor was amazed at his vigor despite the voluminous drinking.

My dad stopped drinking for about one year while I was in graduate school, but the center did not hold. Eventually, he slipped off the knife's edge into the abyss. In the aftermath of his death, his AUD and sincerely held belief that alcohol fully relieved his suffering captured my scientific mind.

I wanted to know why alcohol relieved his pain, why he became so dependent on it—to the point where it was clearly killing him—and why he couldn't stop. Beyond the obvious—the accidents and abuse—I began to find some answers hiding in plain sight, from my own research, just as I moved to Berkeley, California, after six years as faculty in Tucson at the University of Arizona.

Robert Dudley, a Berkeley colleague, has proposed the "drunken monkey" hypothesis to explain the widespread human use of dietary ethanol. His eponymous book was inspired by the death of his own father, which was hastened by AUD. The gist of Dudley's idea is that the ethanol produced by brewer's yeast in fruits can reliably signal ripeness to animals. This sign helps them find the fruit while also increasing consumption rates and the dispersal of swallowed seeds.

Ethanol-containing ripe fruit would only be appealing to those animals able to detoxify it and use it as an energy source. Among the primates, the gorillas, chimps, and humans are by far the most efficient at disarming ethanol as a toxin, thanks to advantageous changes in ethanol detoxification enzymes that evolved about twelve million years ago in a common ancestor.

Not every piece of fruit gets eaten, however, so there are plenty of fruits left for yeast to use and to multiply within. Such a win-win between the plant and yeast is mediated by sugar: yeast uses the sugar in fruit as a source of energy, and the ethanol that the yeast produces kills off competing bacteria and attracts fruit-eating animals, which then spread the seeds. This interaction is another example of how the two sides to natural toxins are ever present in nature.

Ethanol is a toxic larder for yeast, a chemical shield for fruit flies, a potential way that plants attract seed dispersers, and, for us, an energy source and psychoactive drug that binds to $GABA_A$ receptors. But ethanol isn't the only natural toxin that interacts with these receptors. Let's look at some others, including some of the terpenoids we have already discussed.

Shinrin-Yoku and Talking Trees

The slings and arrows of life pushed my father and me to seek refuge in the forest in similar seasons of our lives — in our forties.

Of course, we aren't alone. We know now, thanks largely to the work of scientists in East Asia, that forest bathing, or shinrin-yoku in Japanese — encountering a forest with all five senses — may bring many health benefits, including reduced stress, anger, anxiety, depression, and fatigue and improved immune function, mood, vigor, and sleep patterns.

In a study of ninety-two alcoholics, those who received forest bathing therapy reported significant improvement in symptoms of depression, which a majority of those with AUD suffer, compared with the control group over a nine-day treatment period. This result comports with my own experience with my father, whose depression waxed and waned over life and who practiced his own form of forest bathing, whether it was along the Lester River or in the Sax-Zim Bog.

We don't know for certain why forest bathing has such positive effects on mental health. Its benefits may accrue simply because of the psychological response to exercise, fresh air, and being removed from the many stresses of our modern lives.

However, as Qing Li writes in *The Art and Science of Forest Bathing*, many positive effects of walking in the forest are unique to that setting—they are not achieved, for instance, by walking in a city. There is something special about the essence of the forest.

One idea for the mechanism underlying this effect is that volatiles (e.g., alpha-pinene) emitted by trees may calm us. One line of evidence for this suggestion is that alpha-pinene can hasten sleep in laboratory mice. When given to mice orally, alpha-pinene binds to and activates the same $GABA_A$ receptors in our brains that ethanol and barbiturates and benzodiazepines like diazepam (Valium) bind. Most likely, plants that make alpha-pinene had an evolutionary advantage because this chemical binds to the $GABA_A$ receptor of insects and suppresses herbivory. With my collaborator Jia Huang and our students, I have studied how insects have repeatedly evolved the same changes in the structure of the $GABA_A$ receptor over the past four hundred million years in response to ever-increasing cocktails of terpenoids that bind to them—including terpenoids like thymol, which helps give thyme its characteristic flavor.

Benzodiazepines and other drugs that bind to the same $GABA_A$ receptors are the most widely prescribed sleep medications, including zolpidem (Ambien). These drugs have hypnotic, sedative, and antianxiety effects.

The neurotransmitter GABA was first found to be an inhibitory neurotransmitter in crayfish, suggesting it has ancient origins in animal brains. GABA is also made by plants. They use it much as they use salicylic acid and related compounds, as both a signaling molecule and a toxin that directly suppresses the feeding rates of herbivorous animals, possibly by interfering with the $GABA_A$ receptors.

Like zolpidem, alpha-pinene causes a hypnotic effect, but unlike zolpidem, it has the advantage of not reducing sleep quality in mice. In light of other studies from laboratory animals, the soothing effects of the volatile alpha-pinene-containing essential oils I use in the diffuser in my bedroom may possibly be due to this $GABA_A$ receptor-binding mechanism. All things considered—and the research is still in its early days—these and other experiments suggest that volatile terpenoids like alpha-pinene may be partly responsible for the health benefits of forest bathing.

In the forest bathing literature, and even in marketing copy for essential oil-based products, you'll see the word *phytoncide* used for a plant chemical that can benefit human health. This obscure term was coined by the Russian biologist Boris Tokin, who found that many such chemicals from plants, including volatiles, had antibacterial properties.

Unfortunately, Tokin aligned himself with Lysenkoism—a pseudoscientific and discredited inheritance mechanism that led to massive failures in post–World War II Soviet agriculture. In line with the milieu in which he found himself, Tokin used a faulty for-the-good-of-the-species framework to explain why plants produced these chemicals. This model, however, does not accurately describe how evolution works.

Although his reasoning was incorrect, Tokin correctly understood that plants and microbes produce chemicals that serve as defenses, and he deserves credit for this observation. He also presciently hypothesized that volatiles produced by plants damaged by pests could become chemical messages received by other plants—so-called talking trees. These messages can even be absorbed by other plant species, which can then ramp up their defenses in turn.

We now know that plants use volatile chemicals to "eavesdrop" or at least detect signals from their neighbors and recognize their kin. These volatiles can also be used as signals by the enemies of the plant's enemies.

Parasitoid wasps, for example, use the volatiles emitted by herbivore-injured plants and even the herbivores themselves as an SOS for the wasps to home in on distressed plants. Once they land on the plant, the wasps find the insect hosts they need to complete their life cycle. From the plants' perspective, their relationship with the wasps is like that of the cleaner fish at "cleaning stations" that larger fish visit. While floating over these special spots, the little fish, often wrasses, remove parasites from the larger fish. This arrangement between species is a win-win. In the wasp and plant example, the plant gets a break from herbivory, and the predator gets a meal.

Unlike fish, plants infested with pest insects can't seek out the parasitoid wasps themselves. To get help, they must attract it, often through chemical signals, whether the outcome is pollination, seed dispersal, or attracting the enemies of their own enemies.

It is unclear whether plant volatiles evolved for the *purpose* of attracting wasps. These chemicals could just be a by-product of herbivore-damaged leaves. But in any case, the wasps have certainly learned to take advantage of trees' chemical signals. Homing in on the volatiles emitted by injured plants is one of their go-to strategies for finding host insects. This exchange of goods and services benefits both plant and wasp. So, regardless of the original reason, from the plant's perspective, the enemy of its enemy is its friend.

α-pinene

ambien

Beyond isoprene and rubber, the three terpenoids we've focused on so far in this chapter — alpha-pinene, betulin, and ethanol — seem to have little in common at first blush. However, betulin, like alpha-pinene, binds to $GABA_A$ receptors, can prevent seizures in mice, and has been patented as a potential antianxiety drug. Ethanol also probably binds to $GABA_A$ receptors by mimicking the neurotransmitter GABA itself. Additionally, naturally occurring mutations in the $GABA_A$ receptor genes expressed in our brains are associated with AUD.

It may be just an accident of evolutionary history that these three chemicals are all molecular targets for $GABA_A$ receptors in the brains of animals as different as insects and humans. After all, animals share a nervous system that evolved in the ocean well before they colonized land.

Alternatively, evolution may have favored plants and fungi able to produce chemicals targeting the Achilles' heel that is the animal nervous system — to push the animals away or to draw them in. Plants even deploy some of these chemicals against their own kind. As the Genevan botanist de Candolle said, plants of a given place, competing for the same resources, are perpetually "at war." What he might have meant was that plants are competing with one another, and reducing herbivory and pathogen attack gives them a leg up. But there might also be direct warfare between plants through poisons.

Deadly Drops, Bitter Aloes, and Mad Honey

Shane and I heard a tremendous crash early one morning in the spring of 2017 as a torrential winter storm hit Berkeley. I looked out our south-facing window and saw nothing unusual. Apparently, I was looking in the wrong direction. A few minutes later, I heard a knock at the door. It was my neighbor, who looked upset. She had just returned from dropping her daughter off at school and informed us that a tree had fallen in the parking lot behind our apartment.

I followed her along the sidewalk to find the trunk of a massive blue gum, a species of eucalyptus tree, at least a hundred feet tall and ten feet in diameter, on top of my car. The firefighters and I laughed in disbelief as we walked single file on top of the massive trunk to inspect the remains of my Volkswagen Golf below, which had been split in half and completely crushed. The smell of terpenoids from biodiesel (this was Berkeley, after all) and eucalyptol burned our throats.

I would have been killed instantly if I had been in the car when the tree fell. Although it was an act of God, I couldn't help but think it was a sign, given that I study the natural enemies of plants.

Curious about why a forest of Australian eucalyptus trees was thriving in the East Bay hills, I dug into the biology of these blue gums. The answer is uninteresting; they were planted for lumber and as windbreaks. In my search for answers, I discovered that they have a devious side beyond their car-crushing potential.

Fog drip brings life to the coast redwood forest. Well over half of all the water used by other plant species living below redwoods comes from the water that condenses from the fog around tree leaves and drips down to the understory, according to ecologists (and UC-Berkeley colleagues) Emily Limm and Todd Dawson.

The crowns of blue gums also need fog drip to obtain water. But the ground around these trees is unusually devoid of other plant species, especially outside their native Australia. When I walk under the blue gums here in Oakland or Berkeley, it is eerily quiet and wide open, even in light gaps. An invisible hand may be weeding this garden. Toxins produced by the leaves, bark, and wood of blue gums leach into the moisture delivered by the fog drip and seep into the earth, where, and along with bark, wood, and leaf litter, they can help prevent the growth of other plant species and even beneficial soil bacteria.

The chemicals in the water that leach into the soil around blue gums are terpenoids like eucalyptol and alpha-pinene. Additionally, they contain a dose of phenolics and flavonoids. As a later chapter will show, some of these compounds, at the right dose, may have a protective effect against cardiovascular disease and diabetes when consumed in coffee or tea.

The phenomenon of one plant inhibiting the growth of another through the production of toxins released into the environment is called *allelopathy*, which means "to cause suffering to another." Experiments are required to determine whether allelopathy is actually at play, because many other factors, like shade and water deprivation, can keep the forest floors near large trees relatively free of other plants.

The question is why a blue gum would evolve to prevent the growth of other plant species near it. It could be the same reason, at least in part, that yeast produce high levels of ethanol: to snuff out the competition.

By suppressing the growth of other plants, allelopathy can prevent competition between plants for resources, including water, minerals, and even light. So, unlike the blue gum that hit my car, toxic fog drip may not be an accident of nature. Allelopathy could also be a by-product of the primary role of these allelochemicals as plant defenses against herbivores and pathogens attacking the trees.

As much as I loathed the blue gum that destroyed my car, the eucalyptol wind I encountered as I walked out of the airport in Brisbane the first time I visited Australia instantly gave me a soft spot for these trees.

I *really* like the smell and taste of eucalyptol, and I'm not alone. You probably do too. Eucalyptol is the active ingredient in eucalyptus essential oil. This volatile terpenoid is found in many members of the mint family, like basil, rosemary, sage, sagebrush, and wormwood, and in the daisy, ginger, orchid, mustard, hemp, and laurel families.

Nowadays, eucalyptol is used widely in alternative and complementary medicine. It is deployed as a topical antiseptic, a pill to treat respiratory illnesses, and in aromatherapy, massage oils, and many other products, from candles and soaps to mouthwashes and lozenges.

The widespread use can fool us into thinking that eucalyptol could have only beneficial properties because it was made for us. This belief is the *appeal-to-nature fallacy* — that because something is natural, it is inherently good for us. In Australia, suspected eucalyptol poisoning in children, often from vaporizers, is a leading cause of calls to poison control centers.

Oral consumption of even half an ounce of eucalyptol is extremely dangerous to small children. Seizures are a rare but far more frequent side effect than death. But I don't want to scare you; most cases of accidental consumption by children are not symptomatic.

On the other hand, evidence from Aboriginal Australians and now, randomized, double-blind, placebo-controlled clinical trials shows us that eucalyptol can successfully treat a variety of health conditions, especially upper-respiratory infections like sinusitis, rhinitis, and bronchitis. The successful health treatments include aromatherapy, whose use with eucalyptol even reduced symptoms of dementia in nursing home patients.

Blood flow throughout the brain increases after short periods of eucalyptol inhalation in humans, and eucalyptol produced antianxiety effects in mice when administered orally. Eucalyptol activates the "menthol receptor" TRPM8 discovered by physiologists Diana Bautista (my Berkeley colleague), David Julius, and others and alleviates some kinds of pain while simultaneously inhibiting the "wasabi receptor" TRPA1, which sends to the brain the pain signals arising from heat or compounds in mustards.

Plants don't always use eucalyptol as a poison—and by *use,* I mean favored to do so through evolution by natural selection. The chemical is also produced by orchid flowers, and male orchid bees in Florida, Mexico, Central America, and South America gather eucalyptol along with other scents. As these bees scrape the chemicals from the flowers, packets of orchid pollen become inadvertently attached to their bodies, thereby helping to pollinate the plants when the bees visit another plant of the same species.

The male bees are attracted to the orchids because the flowers are the source of raw materials for the complex perfumes used to entice females into mating. The process is much like how mountain pine beetles use pine trees' own chemicals as a perfume to attract other beetles to the same tree.

Competition for mates mediated by different perfume recipes can even produce new bee species, according to bee biologists Philipp Brand, Santiago Ramírez, and collaborators. They found that both the ability to use new scent sources for perfumes made by males and the origin of new scent receptors in the "noses" of the females (their antennae) can evolve in tandem. A tight evolutionary link can then form between the origin of new scent mixtures made by the males and a novel appreciation for them by the females.

Orchids, over thirty thousand species strong, are more diverse than any other plant family. Vanilla beans are really the fruits of the vanilla orchid, native to Mexico, and the flecks in vanilla ice cream are their innumerable tiny seeds. Vanillin is a phenolic chemical from these seeds and one of my favorites.

The evolution of new volatile chemicals in the orchids may be linked to the plant's ability to receive pollen from another plant or have its pollen moved to another through the movement of a pollinator. In turn, new bouquets are favored as male bees seek out new chemicals for the perfumes they concoct to entice females to mate. The females then evolve the ability to detect and prefer the new scent mixtures. Chemicals that are poisons in one context are co-opted as perfumes in others.

Meanwhile, some plants are nudged by natural selection to put deadly levels of toxins in their nectar. It seems paradoxical that a plant would evolve to poison its pollinators, but not all pollinators are equal in their ability to move pollen to and from a plant.

Evolution may favor plants that are better able to attract the best pollinators and filter out inefficient ones. The most obvious evidence of this theory is found in the palette of colors that evolved in flowers because the various hues appeal to the visual capabilities of the pollinating animals that do the best job at moving pollen around. Blue and violet flowers tend to be bee-pollinated, while orange and red flowers—colors bees can't see well—are often bird-pollinated because most birds can see red well. Orange and red are private color "channels" that plants evolved to advertise to their "subscribers" of a feathered variety.

Once attracted to a flower, the visitor, be it a bat, bee, bird, or bush baby, is rewarded with scents, nectar, or pollen. The plant wins because even if the animal feeds on the pollen, enough of it is moved to another flower of the same species to make up for the loss.

Although the colors, shapes, and fragrances of flowers dominate our perception of them, some plants use toxins to filter out poorly performing pollinators. Some of the best-studied toxins are the phenolic compounds that render the nectar of South African aloes highly bitter.

Honeybees and long-billed sunbirds, both nectar-feeders, reject the nectar of the aloe flowers and are not regularly observed foraging at them. In contrast, two short-billed bird species, the bulbul and white-eye, flock to the flowers, apparently unfazed by the bitter nectar.

From the aloe's point of view, this is a good match: the short-billed birds are far more effective than the sunbirds are at pollinating the many small cup-shaped flowers that line the aloe's stalks. The phenolic compounds found in aloe nectar are also produced by their leaves as defenses against herbivores.

These phenolics are also the same anthraquinones that humans have relied on for thousands of years as a purgative called *bitter aloes*. The toxins come mainly from the latex found in the outer layers of the leaves, not from the polysaccharide-rich gel that is found deeper inside the leaves and that we apply to soothe burns and even drink. The bitter phenolics in the aloe's nectar serve as gatekeepers so that the animals best equipped to move the protein-rich pollen are allowed access and herbivores are blocked from attacking.

Aloes may seem clever, cunning, and intelligent, but they aren't, of course. Such virtues require a brain. These plants are, however, exquisitely adapted to the war of nature because they attract the good pollinators and repel the others by manipulating the animal mind.

Aloes aren't the only plants that evolution has transformed into poisoned chalices. Terpenoids known as grayanotoxins are produced in all the tissues of *Rhododendron* and *Azalea* species, and in some varieties, they are found in the nectar of their beautiful flowers as well. As I'll detail later, these toxins, the active chemicals in "mad honey," were the basis of the first chemical weapons in recorded human history.

Grayanotoxins are potent neurotoxins that bind to the voltage-gated sodium channels of animal nerve cells. These channels are also targeted by the tetrodotoxin in newts and puffer fish, pyrethrins in daisies, aconitine in monkshood or wolfsbane, and batrachotoxin in poison arrow frogs and pitohui birds from Papua New Guinea.

Each of these neurotoxins blocks nerve cells from firing normally. This action leads to heart and nervous system dysfunction, paralysis, and death. Ounce for ounce, these toxins from nature are some of the deadliest.

The grayanotoxins in nectar are far more harmful to northern European honeybees than they are to northern European bumblebees. In Ireland and Great Britain, *Rhododendron ponticum* is invasive, having been introduced from the Iberian Peninsula in the eighteenth century. The honeybees of Ireland and Great Britain are twenty times more likely than native bumblebees are to die after feeding on *R. ponticum* nectar.

R. ponticum is also the source of the nectar used by honeybees from Turkey and the Caucasus to make "mad honey." Notably, the Greek historian Strabo, in *Geographica,* describes how the toxic honey was used to fight Pompey the Great's army in 67 BCE. The Heptacomitae people of Pontus (the Persian kingdom of Mithridates VI Eupator, "the Poison King"), on the southern coast of the Black Sea in what is now Georgia, brought the army to its knees by sneakily feeding them this toxic mixture:

> The Heptacomitae cut down three maniples [600 men] of Pompey's army when they were passing through the mountainous country; for they mixed bowls of the crazing honey which is yielded by the tree-twigs, and placed them in the roads, and then, when the soldiers drank the mixture and lost their senses, they attacked them and easily disposed of them.

What Strabo described is the first written account of the use of a toxin in warfare.

If honeybees are susceptible to grayanotoxin, how could the toxic honey be made from nectar of *R. ponticum?* The honeybee varieties of the region seem to have evolved a resistance to the grayanotoxins and are the pollinators for *R. ponticum* in the Caucasus region.

This poison honey may well be the first documented use of a chemical weapon in human history, but it continues to be used widely in Turkey, where it is called *deli bal,* and by the Gurung people of Nepal, whose national flower is a *Rhododendron.* The Gurung use the honey for food and as a painkiller, but demand in East Asia has soared owing to its purported powers as an aphrodisiac.

On the heels of Pompey's grayanotoxin-stricken army, another terpenoid-based weapon was used at the edges of the Roman republic in battle, albeit in a different way. In 54 BCE, the aged Cativolcus, a king of the Gallic Eburones, whom Julius Caesar's army had conquered, led a rebellion against Rome. The following year, the Roman army avenged the attack with gusto. Cativolcus, too old and feeble to mount a defense and unwilling to surrender, "destroyed himself with the juice of the yew tree," according to Caesar himself.

Yew trees are gymnosperms—like cycads and pines. Yews produce a cocktail of potent toxins that people have used for millennia to poison the tips of arrows, to take their own life, or to poison rivals. There are two main classes of toxins in yew trees. The first class is the taxine alkaloids, which are heart poisons, and the second class is the diterpenoids known as taxanes or taxoids, which affect cell division. Together, these toxins present a formidable barrier to herbivores.

In an effort to find cytotoxic chemicals that might help in fighting cancer cells, the US National Cancer Institute first isolated a taxane called paclitaxel from the Pacific yew tree. The compound showed promise in preventing tumor cells from dividing, and eventually, it was approved for treatment of several types of cancer. Fortunately, because precursors to paclitaxel are found in all yews, the drug can also be made semisynthetically. More commonly known by the brand name Taxol, this remarkable antitumor chemotherapy agent has extended the lives of thousands of people who have cancer.

Although I have covered much ground on terpenoids, we aren't through with them quite yet. The next two chapters focus on two classes of these chemicals that have played an outsize role in both evolution and our own lives. First, I will trace cardiac glycoside heart poisons from the poison arrow tree to monarch butterflies, from foxgloves to heart drugs, and from the skin of toads to preeclampsia. Then, I will wrap up the study of terpenoids with a chapter on how these heart poisons led to the development of hormones from plant toxins, from the Pill to the "Russian secret."

4.

Dogbane and Digitalis

Life that crawled, life that slunk and crept and never closed its eyes. Life that burrowed and scurried, and life so still it was indistinguishable from the ivy stems on which it lay. Birth, life, and death—each took place on the hidden side of a leaf.
— Toni Morrison, *Song of Solomon*

Meadowlands, Milkweeds, and Monarchs

My family moved to the rural township of Toivola, near the hamlet of Meadowlands, Minnesota, the summer before I entered sixth grade. The timing permitted a few months of unfettered exploration before school started.

Conveniently, a thirty-nine-mile abandoned railroad grade ran right in front of our house. I used it to enter the Sax-Zim Bog.

Studded with stunted black spruces, tamaracks, and legions of northern pitcher plants, the taiga-like landscape might as well have been Alaska. The plants weren't alone in this boreal visage. Although secretive, snowshoe hares and timber wolves betrayed themselves in the winter as they laid tracks in the freshly fallen snow.

To walk along the trail is to be transported back in time. When the Laurentide Ice Sheet retreated eleven thousand years ago, it left a depression that formed a meltwater lake called Glacial Lake Upham. The poorly drained lowlands that include the bog are its remnants. Many layers of peat, piled up over thousands of years, hold some of its frigid meltwater to this day. Notoriously cold in the winter, this relic of the Wisconsin Ice Age was a natural refrigerator during the hot, humid summers.

Shortly after we moved, my dad and I traversed the ditch that ran along the road parallel to the trail. As we climbed up the flank of the old railroad grade, both of us noticed a large patch of common milkweeds, plants in the dogbane or Apocynaceae family. Monarch butterflies flitted about the orbs of purple flowers bobbing in the summer wind.

From a distance, we spent a minute or so taking in this perfect portrait of nature. I had a flashback to kindergarten, when our teacher sat at her kidney-shaped table and led a reading circle devoted to the "barfing blue jay" photo series by monarch butterfly biologist Lincoln Brower in *Scientific American*. My classmates shrieked in disapproval as she revealed the last photo, capturing the moment the blue jay vomited just after it had gobbled up the monarch. When the class quieted down, Mrs. Bennett explained that the butterflies contained poisons that caused the bird to vomit.

On hearing this, I had immediately conjured up an image of the ipecac syrup bottle my parents kept in our bathroom at home. They had instructed us to take one-half teaspoon of the stuff if my brother or I had ever eaten or drunk anything poisonous. Its use as an emetic was widespread: in 1984 alone, sixty-eight thousand American preschool children were given ipecac after ingesting something toxic.

It turns out that ipecac is also toxic because the active ingredient is the aptly named alkaloid emetine. *Emesis* means "the act of vomiting." Ironically, although ipecac syrup was used to prevent poisoning after consumption of a toxic chemical, the emetine in ipecac syrup, like the cardiac glycosides in the milkweed sap, are heart poisons themselves—another example of how natural toxins can be double-edged swords.

The tragic death of singer Karen Carpenter in 1983 was caused by "emetine cardiotoxicity due to or as a consequence of anorexia nervosa," according to the autopsy report. Unfortunately, the use of ipecac among people with eating disorders was not uncommon.

Ipecac syrup is derived from the roots of a plant in the coffee family native to Brazil. The plant is called *ipega'kwãi* by the Tupi people of the Atlantic coast. By the mid-seventeenth century, it made its way to Europe. There it was used as an emetic and a treatment for dysentery well into the twentieth century. Nowadays, however, it is a dysentery treatment of last resort, owing to heart toxicity, just as the cardiac glycosides found in milkweeds and foxglove plants are now disfavored as medicines. Of course, these complications hadn't entered my five-year-old mind, and I was mesmerized by the beauty and danger that were rolled into one through the monarch.

As we've seen with other bright, colorful species, the monarch's cinnamon-orange and black-and-white polka-dot wings did not evolve for us to admire; monarchs look the way they do *because* they are poisonous. And their appearance, like a stop sign, sends a strong warning to predators to think twice before attacking. But the monarchs don't make their poisons themselves — they steal them from plants.

After my father and I walked over to the patch of milkweeds, he tore one of the leaves in half. White latex dripped from the leaf. "That's why they call it milkweed," he said. "Don't ever eat it. Heart poisons are in that sap."

As we continued to watch the milkweeds and monarchs over the next month, it was easy to spot the yellow, black, and white striped caterpillars as they hatched from the eggs the butterflies had laid on the leaves, and munched away, oblivious to us, while we inhaled the fragrant perfume wafting from the milkweed blossoms. This tropical scene was out of place in a boreal bog, an Henri Rousseau chef d'œuvre come to life.

The caterpillars of other butterflies and moths that we saw were mostly green, blending in with the vegetation. Some, like the maple leaf-cutter caterpillars that ravaged the trees around our house, even made tiny shelters out of leaves. The monarch caterpillars instead flaunted their gaudy colors in broad daylight.

I knew that birds would vomit if they had eaten a monarch, but I didn't understand why. The butterflies were poisonous, my dad explained, because as caterpillars, they had eaten toxins from the milkweed leaves. The insects then stored the toxins in their bodies all the way through

aspecioside

metamorphosis, from a zebra-striped caterpillar to a chrysalis encircled at the top by a golden diadem, to the familiar brightly colored butterfly.

In entomologist Jim Poff's insect biology course at Saint John's University, I learned about a series of papers published between 1965 and 1968. The articles revealed that the toxins in the monarch were terpenoids called cardiac glycosides. One of the principal toxins in the common milkweeds that my dad and I encountered is aspecioside. The monarchs obtained these heart poisons during their caterpillar stage. But the caterpillars did something even more extraordinary — they concentrated the toxin to levels even higher than those found in the milkweed itself!

In at least fourteen instances, including examples in herbaceous weeds, tropical trees, fireflies, and toads, plants and animals have independently evolved to synthesize cardiac glycosides as a defense mechanism. Plant species containing these heart poisons include crown vetch, foxglove, helle-bore, jute mallow, lily of the valley, oleander, pussy ears (*Kalanchoe* spp.), sea squill, wallflower, and the related suicide tree and tangena tree in the genus *Cerbera*. In Madagascar, the tangena was used in trials of ordeal, wherein the guilt or innocence of the accused was determined when they were forced to consume the tangena nut, which is naturally filled with cardiac glycosides. Upward of 250,000 people, most of whom were enslaved, died from this practice between 1790 and 1863.

Monarch butterflies evolved to become brightly colored to warn preda-tory birds and other predators of the bitter and emetic cardiac glycosides within. Birds that attempt to eat an unpalatable butterfly can then learn to associate the bright colors with the bitter taste of that first bite. Or if the bird goes beyond tasting, eats the insect, and winds up vomiting, it associ-ates the butterfly with danger, just as Pavlov's dogs learned to associate the ring of a bell with food. The multilayered signals that tap into both the innate and the learned responses of predators can be found in the most unlikely places.

The toxic channel of communication that flows between signaler (monarchs) and receiver (birds) therefore enables an extraordinary migra-tion event. Each fall, millions of monarchs migrate up to three thousand miles from their natal habitats in eastern North America, like the Sax-Zim Bog of Minnesota, to the oyamel fir forests of Michoacán, Mexico. It is the same place visited by their great-grandparents or even great-great-grandparents the year before.

The air temperature of these subtropical mountains is just right— usually warm enough that the butterflies won't be killed by frost and cool enough to slow their metabolism so they can make it through the lean months. Like the butterflies, their milkweed host plants were tropical invaders, too, having moved into the northern prairies sometime after the retreat of the Laurentide Ice Sheet. Without the milkweeds, the monarchs cannot breed in the bog; nor would their intergenerational migration take place. But the milkweeds can't uproot themselves and walk to Mexico to escape winter. Instead, they retreat underground, where their toxic, subter-ranean stems rest for the winter and resprout in the spring.

Mother Hutton, Mombasa, and Mutations

The cardiac glycosides that protect milkweeds from most enemies are so named because they affect the functioning of the heart. In 1953, physiologist Hans Jürg Schatzmann discovered that these toxins inhibited the movement of sodium and potassium across the membranes of human cells. Nearly forty years later, in 1997, biochemist Jens Skou won the Nobel Prize in Chemistry for discovering the sodium potassium pump, which the cardiac glycosides bind to in these cells.

Skou found that cardiac glycosides inhibit the pump's ability to move sodium out of cells and potassium into cells, a process that is critical for allowing nerves to fire and heart cells to contract. The heart is where cardiac glycosides convey both their life-extending and life-ending effects.

The bulb of the sea squill, which produces a beautiful flower and is a relative of hyacinths, also produces cardiac glycosides. It is first mentioned in the Papyrus Ebers from around 1550 BCE. The plant has been used medicinally for thousands of years, but its use for treating heart ailments began to come into focus by the first century CE, a fact we've learned from Aulus Cornelius Celsus's *De Medicina*. Celsus ingloriously but accurately recommended it for treating "dropsy," or edema (fluid buildup) in the lungs, which can occur after heart failure: "It is also useful to suck on a boiled squill."

Along these same lines, physician William Withering published a study in 1785 on another plant that produced cardiac glycosides. His research foreshadowed the clinical trial as we know it. He successfully treated 164 patients who had edema with extracts from the leaves of the purple foxglove *Digitalis purpurea*. The report, titled "An Account of the Foxglove and some of Its Medical Uses," led to the widespread use of cardiac glycosides to treat heart ailments.

In his report, Withering recounted how the foxglove was in fact first used as a local folk remedy. The following quote from the report suggests how the seed might have been planted by "an old woman from Shropshire":

In the year 1775 my opinion was asked concerning a family receipt for the cure of the dropsy. I was told that it had long been kept secret by an old woman in Shropshire, who had sometimes made cures after the more regular practitioners had failed.

How exactly do cardiac glycosides work to treat edema? A carefully metered dose inhibits the sodium-potassium pump (sodium pump hereafter) in the heart, as Skou discovered. This inhibition of the sodium pump causes a buildup of sodium in the cells of the heart and in turn drives up calcium levels. High calcium levels strengthen the contractions of the cells; the stronger contractions increase the blood pressure and reduces the heart's rate of contraction. Digitoxin in the purple foxglove and digoxin in the white foxglove were the principal cardiac glycosides used until the mid-twentieth century. The chemical differences between these two drugs lies outside the scope of this book, but their pharmaceutical action is similar.

As digoxin became the drug of choice for many heart ailments, the apocryphal herbalist "Mother Hutton" was invented by the US pharmaceutical firm Parke-Davis (now part of Pfizer) in the late 1920s as the embodiment of the "old woman in Shropshire" mentioned by Withering. Although there is no evidence that Mother Hutton represents an actual person, Withering's account is an example of how traditional knowledge forms the basis for most of modern medicines derived from nature.

Digoxin is still used to modulate blood pressure and to treat heart conditions. In the United States alone, more than 2.6 million prescriptions were written for digoxin in 2019. It remains one of the World Health Organization's essential medicines.

Another important cardiac glycoside like digoxin is ouabain, or g-strophanthin. Ouabain is made by the so-called poison arrow trees (*Acokanthera* and *Strophanthus*) native to East Africa, which, like milkweeds, belong to the dogbane family. The sap from poison arrow trees is used as both an arrow poison and medicine in diverse Indigenous cultures throughout sub-Saharan Africa. The Wilé people of Burkina Faso are said to consider ouabain a gift from paradise to be used as either a poison or a cure, depending on the situation. After people learned of its powers, ouabain became a popular twentieth-century treatment for heart ailments in Europe.

Europeans actually first directly encountered ouabain in 1505, when the Portuguese sacked Mombasa, in what is now Kenya. The Europeans and their ships were met with a flurry of arrows tipped with ouabain. Similarly, Indigenous peoples in Amazonia, the Malay Peninsula, the Indonesian islands, Southwest China, and the Philippines all harvest the cardiac-glycoside-laden latex of several different tropical fig trees to prepare arrow and dart poisons. Again, on the one hand, evolution repeats itself and diverse species re-evolve the same poison, and on the other hand, humans tap into these poisons independently over and over.

In the same vein, other animals besides the monarch and humans have also evolved the ability to use ouabain as a defensive weapon. The maned or crested rat of Africa makes its own ouabain poison from *A. schimperi.* Striped black and white like the cartoon skunk Pepé Le Pew, the rat first chews the bark of this poison arrow tree and then smears the masticate on a special row of spongy hairs that run along the side of its body. The hairs act like wicks to hold the ouabain, an effective chemical shield against dog attacks. The black and white stripes serve as a warning to enemies, just like the monarch's — beware of the striking colors.

Curiously, the monarchs and maned rats seem to be unaffected by the cardiac glycosides they co-opt from plants. It makes sense that plants aren't harmed by cardiac glycosides; after all, they have neither hearts nor sodium pumps. But all animals do have sodium pumps. This raises the question: how do animals that sequester these toxins, as the monarch does, resist them?

I was fortunate enough to have a chance to help answer this question myself, building on over fifty years of research. My work was made possible when evolutionary ecologist Anurag Agrawal and evolutionary biologist Susanne Dobler invited me to collaborate on a research project in 2012.

We decided to use CRISPR gene editing to precisely swap the mutations in the fruit fly's sodium pump gene with those thought to be responsible for cardiac glycoside resistance in the sodium pump gene of the monarch butterfly. In the end, after eight years of work (and many years of failures as we attempted to create the mutants), our team, led by two postdoctoral research fellows in my laboratory, Marianthi "Marianna" Karageorgi and Simon "Niels" Groen, found that the "monarch flies" could withstand cardiac glycoside levels in their diet at concentrations that could probably kill a parade of elephants.

Just like the monarch caterpillars, the larvae of these altered flies incidentally retained some of the cardiac glycosides through their metamorphosis into adults. Three mutations in the sodium pump gene set the evolutionary stage for what Marianna called the "making of the monarch." It was exciting to reconstruct the evolutionary steps taken by gaudy, toxic butterflies over the course of millions of years.

I thought this research was going to be the end of my work on cardiac glycosides, but it wasn't. The monarchs' ability to withstand cardiac glycosides protects them from most birds, but there are a few predators that have pierced the chemical shield.

In 1981, ecologists Linda Fink and the late Lincoln Brower estimated that black-headed grosbeaks and black-backed orioles consumed over two million monarch butterflies at the insect's overwintering sites in Mexico. This was an astonishing number, representing 9 percent of the colony, and turned on its head the tale we'd been told that monarchs were poisonous to birds. How had these birds overcome the toxins?

For one answer, Fink and Brower found that each of the two bird species has a different way of dealing with the toxins in the monarchs. Like the monarchs, black-headed grosbeaks are physiologically insensitive to the heart poisons, but orioles are sensitive to them. This discrepancy is demonstrated in the behavior of the two birds in the wild. The grosbeaks consume the entire body of the monarch, but the orioles carefully dissect it, eating only the muscles and gut contents, not the exoskeleton, which has high cardiac glycoside levels.

Furthermore, orioles routinely release monarchs in the field after capture; grosbeaks rarely do this. The grosbeaks could, somehow, resist the toxins, but the orioles used both taste perception and feeding behaviors to avoid these butterflies.

In 2021, Niels wrote to me to say that he'd dug into the genome sequence of the black-headed grosbeak. The sequence allowed him to assess which genetic changes had evolved in this bird's sodium pump gene. You can guess what he learned. Grosbeaks carry the *same* genetic changes in their sodium pump genes as those carried by some of the toxin-resistant insects that eat milkweeds.

This case study is an incredible example of the way the war of nature evolves blow by blow through escalating adaptations of defense and counterdefense. But there is another important lesson hidden in the details of how these genetic changes to the sodium pump gene work. Cardiac glycosides can bind to the sodium pump in only one spot, the cardiac glycoside binding pocket, and the toxin fits into that spot as a hand fits into a glove. The simplest way for the monarch to evolve resistance would be to block the toxin from binding to that spot. Such blocking can be achieved by only a handful of mutations in the cardiac glycoside binding pocket. The genetic change in the monarch works like this: when the cardiac glycoside molecule tries to bind to the sodium pump, it finds a stitched-up glove.

The next question is why this Achilles' heel of the cardiac glycoside binding pocket evolved in the first place. One potential answer reveals a common strategy in plants, fungi, and microbes. They take advantage of animal bodies and minds by hijacking neurotransmitters and hormones that animals and humans already make, beating us at our own game. From salicylates to GABA and, as we will learn next, cardiac glycosides, a common chemical language is used by many different branches of the tree of life.

5.

Hijacked Hormones

To every thing there is a season,
and a time to every purpose under the heaven:
A time to be born, and a time to die;
a time to plant, and a time to pluck up that which is planted

—ECCLESIASTES 3

Pregnancy and Toad Glands

Lady Sybil, one of my favorite characters in *Downton Abbey*, died during childbirth despite the doctor's attempts to save her life. The cause was preeclampsia—highly elevated blood pressure during pregnancy. On top of this, in real life, one of my own family members gave birth to her baby eleven weeks early, also because of preeclampsia. I then began to better appreciate what real people and their families go through when this serious condition hits.

I did some digging to learn more and stumbled on something that shocked me: there was a possible connection between preeclampsia and my own research on the monarch butterfly. Even more astonishing was how this potential link might explain *why* cardiac glycosides, which evolved in so many plants and even some animals, work so well as toxins. The details are still murky, so bear with me as I try to bring us up to speed on a rapidly changing and controversial area of medicine. By doing so, we will consider a hypothesis explaining why plants and other organisms like toads and fireflies evolved to make cardiac glycosides at all.

Preeclampsia is a major public health problem, affecting up to 8 percent of all pregnancies. High blood pressure, which leads eventually to protein in the urine as the kidneys begin to fail, are two of the main symptoms of preeclampsia. If preeclampsia is left untreated, organ failure, stroke, hemorrhaging, and convulsions can ensue.

Family history, autoimmune disorders, obesity, diabetes, and previous pregnancy with preeclampsia are all risk factors. Although the symptoms emerge later in pregnancy, this potentially dangerous condition of high blood pressure during pregnancy is set in motion, often undetected, in the first trimester. The specific causes of preeclampsia are still unclear.

Strangely, a clue to its causes comes from the curious case of a fifty-six-year-old man with hypertension. He was admitted to the hospital with atrial fibrillation in 1983. To stabilize his heart, he was treated with the typical low dose of digoxin. Because the difference between the therapeutic dose and the lethal dose is narrow for digoxin, the doctors continually measured digoxin levels in his blood as a precaution.

After two months, the doctors halted the digoxin treatment but continued to monitor digoxin levels in his blood. They noticed something strange. The man's digoxin levels kept rising for nine days after treatment had stopped.

The doctors proposed five possible explanations for the still-rising digoxin levels. The most intriguing was the last: "The fifth possibility is that we measured an endogenous digoxin-like substance."

Endogenous means "from within," and the doctors were implying that a digoxin-like chemical might have been produced by the patient's own body as a hormone, like the salicylic acid that our bodies appear to make. As I described earlier, people long thought that salicylic acid and its precursor salicin was made only by plants and bacteria, not mammals like us. But in their report, the doctors cited a paper that also found evidence for "digitalis-like substances" in the plasma of 7 percent of patients who had never taken cardiac-glycoside-based drugs like digoxin.

The presence of glycoside-like substances in people who never took glycosides relates back to preeclampsia because by 1984, researchers had found that levels of endogenous cardiac-glycoside-like hormones in the blood of preeclampsia patients were twice as high as the levels in normal pregnancies.

Soon after this discovery in preeclampsia patients, endogenous cardiac-glycoside-like molecules were isolated from human blood and found to be identical in their chemical structures to ouabain from plants and to marinobufagenin from toad venom. That's right, toad venom.

In hindsight, we might not be so surprised to learn that humans may make cardiac glycosides. For example, some toads produce cardiac glycosides called bufadienolides. These compounds are stored in the parotid glands—those large wartlike bumps positioned above each shoulder, just behind their eyes. Marinobufagenin, one of the most abundant cardiac glycosides made by toads, is so called because it was isolated from *Bufo marinus*, the cane toad.

The use of toad venom as both a medicinal and a poison was widely appreciated in Shakespeare's time, as the First Witch in *Macbeth* portends:

Round about the cauldron go;
In the poison'd entrails throw.
Toad, that under cold stone
Days and nights has thirty-one
Swelter'd venom sleeping got,
Boil thou first i' the charmed pot.

Toad venom is the basis for the ancient Chinese medicine Chan Su, or Senso. In Japan, toad venom is a component of topical anesthetic and heart medicine Kyushin. Toad venom is also used in the purported aphrodisiacs sold as street drugs Lovestone and Rockhard, which have resulted in deaths.

Then there is the nine-year-old boy in Queensland, Australia who, presumably for fun, consumed the eggs of a cane toad he found in a creek. When his mother brought him to the emergency room at the hospital, he was vomiting and drowsy. His skin began to develop cyanosis, a bluish hue from a lack of oxygen.

Underlying the cyanosis was an irregular heartbeat. His heart wasn't pumping blood properly. His condition was so concerning that the doctors administered an intravenous antidote, an antibody isolated from sheep's blood that would bind to the marinobufagenin. Sheep produced the antibody because they had been injected with a modified form of digoxin from foxglove.

As the sheep-derived antibodies dripped into the child's bloodstream, the cane toad's marinobufagenin now in his blood bound to the antibodies, and he recovered. There is a connection here to preeclampsia — scientists are considering whether this same antibody could be used to treat preeclampsia. That's because preeclampsia is associated with higher-than-normal levels of what might be endogenous cardiac glycosides, or chemicals that function like them, in the bloodstream, which in turn may raise blood pressure.

Our bodies do appear to make chemicals that resemble the structures or functions of toxins that other organisms like plants and toads use as chemical defenses against animal predators. However, this supposition is highly controversial, and precise metabolic pathways for synthesis of these chemicals in humans are not fully known. But marinobufagenin could be synthesized in our adrenal glands and placental cells, and ouabain in the adrenal glands and brain.

If we do make these chemicals as hormones in small amounts, then the cardiac-glycoside-binding pocket of most animals is an ideal vulnerability to exploit as a defense strategy. In using cardiac glycosides as a toxin, plants and some small animals like fireflies and toads could be simply taking advantage of an ancient Achilles' heel most animals share, a cardiac-glycoside-binding pocket that is accessible to cardiac glycosides. By overproducing hormones that we ourselves and other large animals produce in minute amounts, a vulnerable plant or small animal can give to its larger animal enemies a big, toxic dose of their own heart hormones. At high doses, cardiac glycosides are heart poisons, but in small doses, they are life-giving hormones and drugs. Another double-edged sword.

A more familiar set of terpenoids, including melatonin and the sex hormones estrogen, progesterone, and testosterone, are also used by humans and other animals to regulate reproduction and the development of their bodies. Not surprisingly, some plants also make molecular mimics of these hormones, which they use against animal attackers. Bringing it full circle, some of the same chemical pathways needed to make cardiac glycosides are used in making sex hormones, too.

Maximilian, Motherhood, and Muscles

Although nobody knew it at the time, an obscure research paper on cardiac glycoside chemistry published in 1944 by chemists Willard Allen and Maximilian Ehrenstein would change history. In the article, they described their success in replacing a carbon atom at position 19 in a cardiac glycoside molecule called k-strophanthin with a hydrogen atom.

Although this result sounds trivial, it was anything but. With these two changes, like alchemists, they transformed k-strophanthin into another terpenoid called 19-norprogesterone, which is one form of progesterone. Progesterone, better known as the pregnancy hormone, is the active chemical in the Pill.

In the mid-twentieth century, there was great interest in developing semisynthetic or fully synthetic routes to progesterone to treat menstrual disorders and prevent miscarriages. When Allen and Ehrenstein injected a small amount of 19-norprogesterone into a rabbit whose ovaries had been removed, its uterus developed a uterine lining consistent with pregnancy. This result suggested that the 19-norprogesterone derived from k-strophanthin mimicked the physiological effects of progesterone.

Progesterone is called the pregnancy hormone because its levels rise just after ovulation, it plays a critical role in preparing the uterine lining for embryo implantation, and it prevents premature contractions that lead to miscarriages. Importantly, it also blocks ovulation.

However, the 19-norprogesterone made by Allen and Ehrenstein was oily and contained several stereoisomers — chemicals whose structures are mirror images, each with different effects. Because this new drug would have to be injected, it wasn't really going to work as a medicinal.

Nonetheless, their discovery laid the foundation for the development of the first oral contraceptive pill. The innovation would eventually transform women's rights and family planning because it enabled people who could become pregnant to decide when and whether to become pregnant.

Other research on hormones was also being conducted around this time. In the early 1940s, several laboratories independently isolated and identified hormones that, like cortisol, are synthesized in the adrenal glands. Such adrenal gland hormones were given to Luftwaffe pilots by the Nazis, who believed they would ameliorate hypoxia. In reality, the hormones did no such thing, but this erroneous belief catalyzed the development of an industrial synthesis process for hormones.

In 1940, chemists Russel Marker and John Krueger took a different approach by turning to plants. They successfully synthesized progesterone in just a handful of steps using diosgenin, a saponin extracted from red trillium, sometimes called birthroot, a plant used in some Native American cultures to aid in childbirth or to treat irregular menstruation. But red trillium was a small plant and a spring ephemeral—the researchers needed larger plants that made more of the precursor chemical. Marker also read that Japanese scientists had isolated diosgenin from the yam *Dioscorea tokoro* in 1936. Knowing that there were relatives of the Japanese yam native to the warm deserts of North America, he packed up his car and drove from Pennsylvania to the southwestern United States and Mexico in search of a related yam to use as a ready source of diosgenin.

Marker found what he was looking for in Mexico, settling on *D. mexicana,* an inedible yam that grows diosgenin-rich tubers weighing over two hundred pounds. Because he found a good source of diosgenin in Mexico, he set up shop there. Marker resigned from his faculty position at Penn State and started the company Syntex (a portmanteau of Synthesis and Mexico) in Mexico. The goal was to use plants as a source of the precursors needed to synthesize steroid hormones that, like cortisone, were sorely sought after as drugs.

In 1945, twenty-six-year-old chemist and Jewish-Austrian refugee Carl Djerassi was hired at Syntex. Building on Marker's work that had started with red trillium and the *D. mexicana* yam, Djerassi would become one of the "Fathers of the Pill."

The Syntex team began using the barbasco yam, *D. composita,* which contains several times more diosgenin than *D. mexicana* does. Historically, the barbasco yam's medicinal properties were widely used by the Indigenous people of Mexico to induce abortions and to stun and harvest fish. *Barbasco,* it should be noted, is a Spanish term for "plants with fish-poisoning properties."

Just a year after arriving at Syntex, Djerassi and Hungarian Mexican chemist George Rosenkrantz and the rest of their team successfully synthesized estrogen, cortisone, and testosterone from diosgenin. They next set their sights on development of a semisynthetic progesterone.

As a starting point, Djerassi was inspired by Allen and Ehrenstein's experiments. He sought to repeat them, but instead of synthesizing 19-norprogesterone from k-strophanthin, he used the more readily available diosgenin.

It didn't take long for Djerassi and his team to successfully synthesize 19-norprogesterone from diosgenin. It was an important step, but it didn't quite get them the drug they were looking for: Like the product Allen and Ehrenstein had synthesized, Djerassi's 19-norprogesterone was oily and insoluble in water. Because of these properties, it would never work as a pill, as it wouldn't be absorbed into the bloodstream.

Instead, the team turned to 19-nortestosterone as a starting point. Like Allen and Ehrenstein, they swapped the carbon at position 19 with a hydrogen. The laboratory notebook of twenty-six-year-old Luis Miramontes, a National Autonomous University of Mexico undergraduate student who was working under Djerassi at the time, reveals when Miramontes first synthesized pure 19-nor-17-ethynyltestosterone, found in most versions of the Pill used today: October 15, 1951. I like telling this story to my undergraduate researchers—given the right training environment, anybody has the potential to make an important scientific discovery.

The synthesized hormone was tested on animals and found to be the most effective oral contraceptive drug known. The team filed for a patent, and in 1952 Djerassi, Miramontes, and the rest of the team officially published their findings. The drug became known as norethindrone, initially licensed by Parke-Davis in the United States.

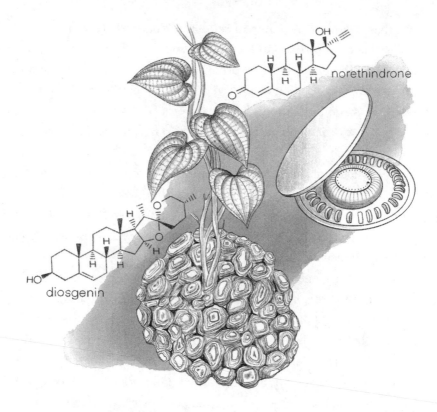

Great minds often think alike. In 1953, Frank Colton, a chemist working for G. D. Searle & Co., filed a patent for a nearly identical chemical, an isomer of norethindrone called norethynodrel. When the substance was swallowed, the stomach's acidic environment converted Colton's drug into norethindrone.

Both drugs worked to prevent ovulation, but the Searle drug made it to market first, receiving FDA approval in 1960. Although the team in Mexico discovered norethindrone first, they were commercially scooped by Big Pharma.

Still, Syntex kept at it and won FDA approval two years later, and by 1964, several large pharmaceutical companies became interested in its version. Ortho (then a division of Johnson & Johnson), Syntex, and Parke-Davis all ended up using Syntex's norethindrone in the most widely used version of the Pill. Soybeans have now replaced yams as a basis for the synthesis of these critically important hormone analogs.

Yet there is still interest in yam extracts as "natural" remedies for various gynecological disorders. Indeed, if you search the internet for the word *diosgenin*, you will find a variety of products containing it for sale, including creams and dietary supplements often derived from wild yam, fenugreek, or soy, three distantly related plants. However, there is no evidence that diosgenin is converted by our own bodies into hormones that can mimic the effects of progesterone.

In 1935, entomologist and physiologist Gottfried Fraenkel teased out one of the functions of diosgenin in plants. Using flies that grew from the piece of meat he had accidentally left in his laboratory in London, Fraenkel discovered the hormone 20-hydroxyecdysone, or 20E, which insects require to shed their exoskeletons as they grow.

Some plants endogenously synthesize 20E from diosgenin and use it as a way to punish herbivorous insects, because if the insects eat it, in many species, their molting and reproduction are impaired. Thousands of plant species can imitate this hormone, and in the most extreme cases, these chemicals make up to 3 percent of the plant's weight.

The plant version of 20E is aggressively marketed to (and used by) athletes and bodybuilders because many believe it works to increase muscle mass and strength. In fact, 20E may have been the main ingredient in the "Russian secret" of the 1980s, which is, for now, a legal supplement. The rationale for 20E use by athletes is that these steroids from plants are natural anabolic agents that are unlike anabolic-androgenic steroids (commonly called anabolic steroids) in a few important ways.

Does 20E really trigger the growth of new muscle? To find out, the World Anti-Doping Agency commissioned a randomized, double-blind, placebo-controlled trial to test whether 20E affects muscle growth and performance. The sample size was small; only forty men recruited completed the ten-week study. Yet, 20E supplementation clearly increased muscle mass and performance. In all, the effects of 20E were even more potent than those reported for the anabolic steroid methandienone, which is banned in sports.

However, a word of caution is in order. More research is needed before any decisions are made on banning 20E. The hormone 20E might work in the following way to increase muscle mass and strength: Although anabolic steroids bind to the androgen receptor to increase muscle mass, 20E binds to the estrogen receptor beta. When the estrogen receptor beta was experimentally activated, as 20E is thought to do, in rats, skeletal muscle fibers of the large muscles increased dramatically in size. So, 20E may bind to the estrogen receptor beta and trigger the growth of skeletal muscle. These experimental results are further evidence of 20E's muscle-building properties.

There are some potential downsides to 20E, because skeletal muscles may not be the only tissues to grow in response to this hormone when people take it. There is concern that 20E may also bind to mineralocorticoid receptors and cause overgrowth of the kidneys and subsequent kidney disease, as it has in lab animals.

The evolution of 20E—an insect-hormone-mimicking steroid made by plants—had absolutely nothing to do with us, despite how 20E affects our bodies. High levels of plant-made 20E and precursors like diosgenin probably evolved to defend against arthropods, not to help gymnasts gain an advantage on the parallel bars. Again, in this situation, plants are making animal hormone mimics that beat animals at their own game. We humans have simply intercepted these analogs and used them as tools.

Where does this circuitous journey from Maximilian Ehrenstein, Mexican yams, motherhood, and muscle mass lead us? A line can be drawn from the single most influential drug ever invented, the Pill, to the herbivore-repelling, diosgenin-laden tubers of wild yams in Mexico, the red trillium birthroot of the Native Americans, and the ouabain in the poison-tipped arrows that the Giriama people used to defend against the Portuguese invaders in 1505.

Just as we use toxins made by plants to augment our hormones, so do some animals. The following remarkable story will connect us to the next three chapters, which all focus on alkaloids.

Montane Voles and Melatonin

My family's house that sat next to the Sax-Zim Bog was surrounded on two sides by an old field used for making hay. Shortly after moving there, my brother and I dug a few holes at its edge just to see what animals we might catch. The most commonly captured in our pitfall traps were meadow voles, which are mouselike rodents related to hamsters. Although I didn't know it then, the reproductive cycle of the voles was under the influence of toxins produced by grasses, their favorite food.

Thirty-five years later, voles entered my life again as I walked along the county road in Vermont in late fall. The plow had just come through after an early snowstorm, and the snowbanks were several feet high. I was just about to turn up our driveway when I noticed a small rodent frantically running from one side of the icy road to the other, trapped by the snowbanks. Wearing thick gloves—don't try this at home, for many different reasons—I gently grabbed the vole to help it get over the snowbank. As expected, it bit my glove, and I quickly released it in a brush pile. I wondered what the vole was doing out in the middle of a snowstorm, and I recalled that voles don't hibernate. This one must have been caught off guard by the sudden snowstorm.

The lesson is that the seasons don't always cooperate with the calendar; fall can come early, and spring can come late. The next sabbatical stop after Shane and I left Vermont was Santa Fe, New Mexico. Although we hoped for a sunny day when we woke up on our first morning there in early March, a fresh layer of snow had fallen on the junipers and piñons.

In addition to being active all winter long, voles live short lives, often for only one year. From the perspective of the vole, then, a more accurate predictor of the food supply than day length or snow levels is the ebb and flow of the actual vegetation that it eats. At high latitudes and elevations with short growing seasons, animals like voles must be able to predict the seasons.

It turns out that the reproductive cycles of voles, including the montane voles that lived in the grassy meadows around the house we rented in Santa

Fe, are closely tied to the amount of grass available. The availability of grass in turn depends on the unpredictable weather in the Rocky Mountains. In some years, an early spring thaw spurs a burst of growth in salt grass, one of the voles' favorites. When these freak thaws occur, the voles are ready to pounce and can squeeze out an early litter of pups by taking advantage of the food after a long winter living on dead grass under the snow.

A big question was how the montane voles could reproduce so quickly in response to these periodic green-ups in early spring. It is as if there is a hidden signal in nature telling them to start producing pups at just the right time — one year it might be late February, the next year late March.

These boom-and-bust cycles are found in other rodents too. The Norwegian language even has the words to describe a mouse year (*museår*) or a lemming year (*lemenår*). Some of these alternating cycles for voles are explained by concomitant increases and decreases in the amount of vegetation and, incredibly, in the toxins that the grasses make.

Grasses account for about twelve thousand species worldwide. We rely on grasses more than we do for any other plant family for our daily nutritional needs. Barley, corn, millet, oats, rice, sorghum, sugarcane, and wheat collectively provide over half of our daily caloric intake. Other animals like to eat grasses too, and as a result, evolution has endowed these plants with antiherbivore toxins that are abundant in their leaves, stems, and roots.

One of the most effective chemical protectants produced by grasses is actually a toxin precursor, or protoxin, called 2,4-dihydroxy-7-methoxy-2H-1,4-benzoxazin-3(4H)-one, or DIMBOA. That's a mouthful, but DIMBOA is an indole, a class of chemicals derived from the amino acid tryptophan. Indoles in turn fall into the alkaloid class.

Grasses use DIMBOA to defend against a diverse group of attackers. Its categorization as a protoxin means that when an animal wounds the plant, an enzyme already produced by the plant converts the DIMBOA to the active toxin 6-MBOA. When herbivores ingest 6-MBOA, most of them find it bitter and unpalatable. However, some herbivores, like voles, have evolved ways of overcoming the distaste and negative effects of 6-MBOA.

Although 6-MBOA provides an excellent defense for the grasses, some insects and nematode worms have also evolved a way around its toxic effects and have found a niche with grasses. These grass-feeding herbivores can detect the 6-MBOA and use it as a marker to find grasses, because only grasses and their close relatives make this chemical. In other words, the 6-MBOA betrays the identity of the plants that these specialized animals need to complete their life cycles.

Montane voles go even further. In 1981, researchers discovered that 6-MBOA from the grasses eaten by the voles can trigger reproduction in these mammals. A close look at the chemical structure of 6-MBOA reveals that it is an analog of the hormone melatonin.

Melatonin is an essential hormone made by the pineal gland deep in the brains of all mammals. It binds to melatonin receptors in the brain and is a master regulator of the circadian rhythm, the wake-sleep cycle, and controls when animals begin and cease to reproduce. Millions of people use melatonin as a sleep aid because it can override the body's response to day length. Melatonin is made by plants too, seemingly in response to environmental stress. This is yet another case of the same hormone being made by both plants and animals.

But humans aren't the only animals dosing themselves with melatonin—or at least chemicals like it. Voles given 6-MBOA had thicker uterine linings, heavier ovaries, and more mature follicles than do those given saline. Even more incredibly, 6-MBOA skewed the sex ratio of their litters toward female pups over male pups.

Researchers have theorized that 6-MBOA stimulates melatonin production in the pineal glands of the voles or may even bind to melatonin receptors in the brain. These actions would in turn influence other hormones involved in reproduction.

How might the connection between 6-MBOA and reproduction work in nature? The levels of its precursor, DIMBOA, in the grasses are tied to their abundance. Early in the growing season, there is little DIMBOA around, but levels quickly increase as the grasses grow. Consequently, 6-MBOA is a faithful marker of the amount of food that will be available to the voles when they begin to nurse their pups and a much better indicator than day length, which is used by most animals as a sign to begin reproducing.

Because these animals' reproductive cycle is so closely tied to the levels of 6-MBOA, when the grasses grow during the freak snowmelt periods, the voles go along for the ride and rapidly produce another litter. If they were to rely on day length alone to trigger reproduction, they couldn't take advantage of the early snowmelt before another snowstorm hit or before their short lives ran their course. Through evolution, these rodents gained the ability to use 6-MBOA as a way to enhance reproduction.

Although the process unfolds innately in voles and by choice when we use the Pill, these rodents, like us, use toxins from plants not intended for them to control their reproduction. However, 6-MBOA promotes rather than inhibits reproduction in these voles, so the outcome is the opposite of that of the Pill.

Remarkably, although voles don't consciously seek it out, either, another kind of toxin produced by grasses influences reproduction in voles—this one by suppressing reproduction, much like the Pill. Given that they live such short lives, it might also be advantageous if voles could shut down their reproduction at the right time in the fall, so that they aren't nursing pups when the first snow falls. The same scientists who discovered that 6-MBOA triggers reproduction determined that a different set of toxins is used by the voles to do just that—to deter it.

In Utah, where the montane vole lives on the edges of the Great Salt Lake, the buildup of phenolic toxins at the end of the growing season in the salt grass inhibits ovulation. This buildup occurs just as the grasses are about to die back in the fall. In other words, the voles stop reproducing thanks to the accumulation of these phenolics at just the right time before their food supply runs out.

This strategy to stop reproduction before winter arrives is adaptive because the voles aren't nursing pups at the wrong time. What's more, they are better equipped to squeeze in an extra litter in the unpredictable early spring thaws. To achieve the best timing for reproduction, the voles must pull the throttle back on their reproduction as the main growing season ends and the fall and winter come. The grasses are the best cues for the triggering (through 6-MBOA) and the dampening (through phenolics) of reproduction because the voles' growth is tuned to plant growth. The same remarkable relationship between these two plant toxins and reproduction applies to montane voles as well.

The phenolics that cause voles to stop reproducing are, incredibly, the same chlorogenic acids in the toxic fog drip of the eucalyptus that kills neighboring plants! Chlorogenic acids, found in many plants like grasses, reach their highest concentrations after flowering and fruiting, for example, just as the grasses finish growing for the season. I'll discuss these phenolics in more detail later, in a later chapter.

This chapter illuminates three simple observations. First, some hormones are produced by animals, humans, and plants. Second, plants and even some animals that have adapted to resist these hormones can overproduce the hormones to protect themselves from attack. And finally, both humans and other animals can turn the tables on plants and use pharmacologically active chemicals in the plants they eat to regulate important bodily processes like reproduction.

Like 6-MBOA and the emetine in ipecac syrup mentioned earlier, many chemical shields produced by plants, fungi, and animals are known as alkaloids. The term *alkaloid* is derived from *al-qili*, the Arabic word for "ashes from plants." It is among the alkaloids that the chemical subterfuge reaches its zenith.

More than any other chemical class, the alkaloids have changed our world by penetrating the innermost sanctums of our minds and bodies, for better and for worse. Alkaloids predominate among the medicines from nature and the drugs of addiction. Before we understand why, we first need to trace their origins. The examination starts with the smell of death.

6.

Abiding Alkaloids

And the whole earth was of one language, and of one speech.

—Genesis

Blowflies and the West Texas Wind

As Catherine de' Medici's perfumed leather gloves foretold, humans go to great lengths to avoid the smell of death. Sometimes, as I learned, there is just no escape.

After I moved out of the temporary Berkeley faculty apartments in November 2017 (without a car, thanks to that eucalyptus tree), Shane and I set up our new little home in Oakland. Over the winter break, we began to explore the local hiking trails in the Oakland Hills. On Christmas Eve, he and I moved through the foggy air that was spiced by the California bay trees. I couldn't hide my anxiety as I obsessively crushed some leaves, inhaling their peppery aroma. I knew that the tree was used by the Yuki people of the Klamath River basin to treat headaches—but the terpenoid umbellulone found in its leaves could also cause headaches by triggering the "wasabi receptor."

I was preoccupied because my dad hadn't returned my texts or phone calls in days. At that point, I'd lost count of how many. Maybe his cell phone was lost, maybe he decided to stop communicating, or maybe it was because of something worse.

My brother, my mother (my dad's ex-wife), and other relatives who checked in on him reported the same radio silence. Although nearly completely estranged from us by then, he responded to nearly every phone call or text. We were all worried, but it was hard to know what to do.

A few days earlier, we had decided it was time to intervene and call the RV park office to check in with the personnel and express our concern. When nobody picked up, I left a voicemail and waited.

Every year on Christmas Day, I make cinnamon buns using a recipe I like from the *New York Times*. The dough must be started the day before so that it can rise and set. The last item on my to-do list that Christmas Eve after our hike was to roll out the dough, paint it with melted butter, and sprinkle an obscene amount of sugar and cinnamon on top.

Me being me, I went down a rabbit hole that night, to focus my ruminating mind, and read as much as I could about the origins of cinnamon. These days, the spice is mostly produced from the bark of the *Cinnamomum* trees in South Asia, either from Ceylon (also known as true cinnamon, or *C. verum*) or Chinese cinnamon (known as Chinese cassia, or *C. cassia*). The trees belong to the same family as the California bays. Cinnamaldehyde gives cinnamon its prickly oomph, and there is nothing quite like it. The chemical activates the same capsaicin receptor found in the nerve endings that line our mouths. Capsaicin produces the heat we feel when we eat chilies.

I set my alarm to wake up early on Christmas Day to put the buns in the oven. A phone call from a Texas area code woke me instead.

I sat up on the edge of the bed and answered. The RV park attendant had bad news. The county sheriff had done a wellness check on my dad after my uncle had gotten through two days before. They'd found his body on the floor of his fifth-wheel trailer. He had died in his trailer, apparently alone.

Shocked but not surprised at the news, my brother and I made immediate arrangements to meet in Texas and get our father's affairs in order. We flew to Dallas Fort Worth, picked up the rental car, and headed west. A few hours later, we arrived at the RV park and walked toward the fifth-wheel trailer through the punishing West Texas wind that now delivered the smell of death.

My most fervent desire at that moment was to run back to the car and flee, as I'd been doing my whole adult life, but I knew that this time was different. He wasn't there, nor were his weapons. His twenty-plus guns, thousands of rounds of ammunition, and knives, including one blade lashed to the end of a long walking stick, had all been removed by the county sheriff for safekeeping. One of his neighbors approached us and warned that there might be booby traps and hidden weapons in the walls. My brother and I shook our heads in disbelief.

While his passing was tragic, it was also a relief. My father was obsessed with guns. So much so that he kept a derringer in his bathrobe pocket, a practice that had always frightened me. But in the months before his death, I had become increasingly anxious because of his gun fixation.

When I was around twelve, he had begun morphing from a gentle outdoorsman into an angry, paranoid, and gun-obsessed man. First it was Rush Limbaugh's voice on his car radio, and then my father would parrot how the government was taking his money and giving it to people who didn't work. Then it was the *Guns & Ammo* magazine subscription, the National Rifle Association membership, and the shotgun placed under my bed, just in case "we" needed it. Eventually, a target was set up in the yard so he could unload into it bullets from his newly acquired SKS Chinese assault rifle on his days off and after work. I didn't like firing it, but occasionally I stood next to him as he did, close enough to hear him murmur, "That is his head"—making sure I knew that the target was a proxy for somebody who had crossed him. It was shocking and deeply disappointing to hear those words. My respect for him evaporated.

How did it all begin? It is impossible to know, but a story he shared just once gave me a clue. When he was a boy, he watched as his intoxicated father threatened his mother with a pistol in the kitchen. He told me that the experience terrified him.

As far as I am aware, he had never been diagnosed with AUD. He did mention that US Department of Veterans Affairs staff had once knocked on his door of his fifth-wheel trailer to see if he was interested in being involved in a study on AUD and post-traumatic stress disorder (PTSD). He told me that he was initially keen to help but, after some thought, decided it was just a ploy to lure him in and force him to stay at the hospital, so he declined. In Texas, his weapons were his legal right to keep as long as he didn't threaten anybody, enter a mental health institution, or have a diagnosis of mental illness, which included drug use disorders like AUD.

The RV park manager had padlocked the fifth wheel's door after his remains were removed. As we waited for the manager to arrive, ever the biologist, I noticed blowflies tracking upwind before they pivoted to fly around the trailer as if they were in a drag race.

I was strangely relieved by the presence of the blowflies. My salve in all the seasons of my life was to get lost in whatever page of the book of nature was open, as macabre as the setting was. The blowflies were just doing their thing, I thought, as my mind trailed off into an intellectual refuge.

Teetering on the edge, I told myself I'd encountered the smell of death many times before. Cadaverine was cadaverine, putrescine was putrescine, and dimethyl trisulfide was dimethyl trisulfide, regardless of what had produced them.

I recalled how I'd even live-captured juvenile Galápagos hawks that were feeding on a maggot-infested sea lion corpse on the beach. As awful as the smell was, the DNA samples from those hawks and their lice were essential to my dissertation research. The same smell brought both the flies and me to the carcass, but for different reasons.

Still, I could not deny the awful truth. I knew which smell receptors in my nose were binding to molecules produced by bacteria that had been breaking down the proteins in my father's once-living body. Knowing the specific receptors didn't provide much comfort.

Though it was normally my escape, just once I wished this profusion of cold, hard biology would cease. Then I reminded myself that I had a fulfilling professional life, thanks to my ability to use biological research as an escape. This escape was better than the alternative, an obsession with ethanol, perhaps other drugs of abuse, and guns. Obsessiveness of a certain kind had broken my father, but an obsession of a different kind is what saved me, time and again.

The last time I'd spoken to my dad was Thanksgiving Day. As usual, I had limited the conversation to ten minutes by starting a timer. It was both one of the more concerning conversations we'd had and one of the best since he'd run away from us. He began by asking me if I'd "been to space."

"You mean, am I an astronaut?" I replied, thinking maybe he had become confused about what kind of scientist I was and was losing it. I was right about the latter, but not the former.

"No," he said. "Have you been to space using your mind to travel there?"

I told him I hadn't but that I wanted to hear more. A lump began to form in my throat. He said that he had been to space and although it was "dark and cold," it was also the place where he could help the people he loved by changing their minds so that they would do the right things — the things he thought they should do.

I changed the topic. Because it was Thanksgiving, I redirected the conversation toward gratitude and told him some of my truths. I credit the twelve-step support group Al-Anon, which is for friends and families of those with AUD, and individual counseling for changing my perspective toward him. I told him that he had been a great father when I was younger, that I was a biologist because of him. I added that I was a good human, in part, because of him and that many of my successes in my professional life flowed through the natural history knowledge and work ethic he had imparted to me. All the negative things were true, too, at the same time, of course, but I didn't mention them. I was grateful for the positives. The positives aren't always there in families affected by chronic drug use disorders.

He began to weep after I shared my feelings about him. I ended the call by telling him that I loved him. He said the same, and we hung up. Although it was a disturbing conversation in some ways, I am grateful for having been able to share some gratitude near the end of his life. More than that, I am lucky that my young childhood wasn't worse and that I wasn't abused or overly traumatized through his behaviors.

Meanwhile, David (a pseudonym), the RV park attendant, rolled up in his ATV wearing a red Make America Great Again (MAGA) hat and a camouflage jacket. Smiling, he shook our hands and introduced himself as both a "Trumper" and a Gulf War veteran.

David said he had helped watch over our dad, whom he remembered as a little "out there" but also as a veteran and a good man. David then said, "Your dad told me about how proud he was of both of you, of your successes in life, and how much he loved you." Tears began to well up in everyone's eyes. I couldn't figure out if it was because of the smell, the wind, or the sweetness of that sentiment. Maybe it was all three.

David noted that he had found a glass stein about half filled with beer sitting on our father's kitchen table, along with a bowl of pretzels. "Your dad liked his beer, as you probably know," he quipped. Did I ever.

As a kid, I would ride with my dad to the liquor store on London Road in Duluth, where we exchanged his cases of empty lager bottles for new ones. He used to keep his stein in the refrigerator during the day, and after getting home from work—selling used cars or furniture—he kept it filled until he went to bed.

I once asked him where the lines of bubbles on the sides of the glass came from. He explained that carbon dioxide was a by-product of yeast's making ethanol from sugar. When a bottle is uncapped, he said, carbon dioxide bubbles come out of solution as the pressure is suddenly reduced. Slight imperfections in the glass stein triggered this off-gassing in those places, he said.

Every now and then, I would sneak a tiny taste but found beer to be unbelievably bitter. I didn't understand why he drank it. "It's my medicine," he would say, in a way that seemed logical and ended the conversation even though I had more questions, as always.

Before David unlocked the door, he cautioned that the smell of death in the trailer was overpowering. He suggested that we borrow his unused army-issued hazmat suit and gas mask equipped with charcoal filters from his tour of duty.

I announced that I was *not* going in. My brother, who'd been in the air force, was then in the construction business and was just made for situations like this. He donned the hazmat suit and gas mask.

As he climbed the stairs to the trailer, he looked like he belonged in a bomb squad. He opened the door and crossed the threshold. The blowflies followed him in.

My brother reappeared a few minutes later. I was relieved. There were no booby traps or weapons in the walls after all. As he walked down the stairs toward me, he held two pill vials. Without a word, my brother handed me the vials as his Darth Vader breaths moved in and out of the mask. My heart sank as I read the labels. In the vials were pills made from alkaloids, collectively called opioids, that target the nervous system. The precursor chemical that opium poppies use to make natural opioids (like morphine) and that we use to make synthetic ones (like fentanyl) is the alkaloid piperidine — the same compounds found in the eastern white pine needles of the boutonniere I wore on my wedding day. Piperidine forms the basis of many alkaloids. Its precursor chemicals, like cadaverine and putrescine, are also found in the smell of death.

The Smell of Death

Although most people have not had the misfortune of smelling their *own* dead, we have all encountered the smell of death: the stench of roadkill, the remains of the rat in the wall, the bloated fish on the beach, the old carcass hidden in the tall grass, or the slimy head of lettuce in the crisper. So powerfully repugnant is the smell of death that even rats held in captivity bury the bodies of dead rats because of it.

Decomposing animal, fungal, and plant tissues produce a bevy of putrid chemicals. The aptly named cadaverine and putrescine are among the most memorable. These so-called biogenic amines are by-products of decomposition, during which amino acids are broken down by bacteria.

In addition to biogenic amines, volatile sulfur compounds like dimethyl disulfide, dimethyl trisulfide, and methyl mercaptan, which are produced by bacteria from sulfur-containing amino acids, also contribute to the smell of death. Finally, we can't forget about indole and skatole, produced from the amino acid tryptophan during bacterial decomposition.

Humans avoid the smell of death like the plague — quite literally, in the case of the perfumed beaks filled with aromatic herbs and used by seventeenth-century plague doctors in Europe. Our disdain for these smells is triggered by a particular set of odorant receptors in our noses. Flies have their own odorant receptors that underlie their attraction to, rather than revulsion by, the same odors, up to a point.

However tidy our relationship with these chemicals of death may seem, it is not so simple. Even these chemicals have two sides.

Cadaverine, putrescine, and the related spermidine — so named because it was first found in human semen — are found not only in decomposing tissues but also in fish and fermented food and drink. Sometimes the levels are high enough to be toxic.

Skatole lives up to its name, producing a ghastly odor when volatilized. The chemical can be toxic if too much enters the lungs. Skatole is also made by some plants and, at low concentrations, helps give the essential oils of jasmine and orange blossom their intoxicating smell. Now we can see why, in the sixteenth century, the word *intoxicated* changed meaning from "poisoned" to "drunk." Skatole has two sides too.

Even more surprising, our (living) bodies produce endogenous cadaverine, putrescine, and spermidine because these molecules serve critical roles in our cells. Spermidine is particularly interesting. Adding it to the diets of laboratory animal models extended their life spans by 15 to 30 percent. In human cells bathed in spermidine, aging also slowed. However, we are only just beginning to understand the potential mechanisms.

One effect of spermidine is that it helps keep our genes switched off. As cells age, more and more genes get turned on willy-nilly. The firing of too many genes can be problematic. Spermidine seems to keep older cells operating as they did when they were younger by reversing this trend to some degree.

Small clinical trials show that increased spermidine in the diet improves cognitive function in patients with a kind of dementia. Spermidine may also enhance the removal of damaged cells, including those containing the amyloid-beta plaques that can accumulate in the brains of people with Alzheimer's disease.

Although our bodies make spermidine, much of it comes from the diet. Should we all be consuming more of this chemical? It is not clear. But like nearly all the chemicals covered in this book, spermidine can have both toxic and life-enhancing effects. So, if we supplement our diets with spermidine, we should be careful to limit our intake. As always, the appeal-to-nature fallacy applies—these chemicals didn't evolve to benefit us.

Cadaverine, putrescine, and spermidine also form the backbones of many alkaloids. From caffeine to capsaicin, mescaline to morphine, and piperidine to psilocybin, these alkaloids are synthesized from biogenic amines and end up targeting the animal nervous system as toxins. Nevertheless, we also seek them out for their medicinal and intoxicating properties.

The lesson here is that one organism's trash is another's treasure. Bacterial decomposition products that we find repugnant and that flies are attracted to were turned on their heads by plants and fungi. They use them as chemical starting points for the synthesis of many of the alkaloids we use as drugs. A common chemistry of life underlies it all.

The smell of death wafted into the London air through an open laboratory window at University College in 1933. A female blowfly homed in on the source as she zigzagged through the fetid plume. Finally, she landed on a piece of rotting meat from Gottfried Fraenkel's forgotten lunch. Yes, the same Fraenkel who discovered 20E.

Fraenkel eventually found the source of the stench in his lab, but by that point, the piece of meat was covered in maggots. Given how easy it was to culture blowflies, Fraenkel decided to use them for his next experiments. A few months later, thanks to that blowfly, Fraenkel discovered 20E, the so-called molting hormone that triggers metamorphosis in all insects.

Fraenkel then turned his sights on a different problem. Over half of all insect species feed only on the tissues of living plants—they are herbivores. Yet the vast majority of herbivorous insect species are picky eaters. Each will attack only one or a handful of closely related plants.

For example, monarch butterflies lay eggs only on milkweeds, and cabbage white butterflies only on mustards, but not vice versa. It isn't just the adults that are so choosy either: a monarch caterpillar won't eat a mustard leaf, and a cabbage white caterpillar won't eat a milkweed leaf. Fraenkel wanted to understand why these insects were so specialized on different toxic plants.

A big hint came from researchers who found that they could persuade specialized herbivores to eat the leaf of a plant they normally wouldn't eat by applying an extract of chemicals made from washing the leaves of their normal host plants. Fraenkel had two hypotheses to explain these preferences. One hinged on differences in nutritional needs between herbivores, and the other hinged on differences in the types of toxins present in the plants.

The first was rather intuitive: the specific nutritional needs of one insect species might only be provided by one set of closely related plant species. So, for instance, Fraenkel supposed that there must be something missing from the mustards for the monarch caterpillars and from the milkweeds for the cabbage white caterpillars.

This theory would have been a simple explanation. Unfortunately, it was wrong. Fraenkel found that all leaves are more or less nutritionally equivalent, whether from a potato, a palm, or a pine. From this finding, he concluded that there was no reason a monarch caterpillar couldn't live on mustards instead of milkweeds, if only you could get the insect to eat them.

Fraenkel then tested the idea that adaptation to the diverse "secondary compounds" made by plants, which include the toxins, could explain the narrow ranges of each insect. He discovered that the patterns of host plant specialization across herbivorous insects was indeed due to insects' different tastes and tolerances for different plant chemicals, some of which are poisons. Each group of insects became adapted to the toxins made by a particular group of plants. Eventually, the insects used the toxins as a signpost to find and eat the right host plants on which they were well adapted. Monarch caterpillars ate only milkweeds and other members of the dogbane family, and anise swallowtail caterpillars only ate plants like anise and other members of the dill family. Fraenkel's findings were published in a 1959 paper titled "The Raison d'Être of Secondary Plant Substances."

But how were toxins that evolved to dissuade attackers eventually turned on their heads by specialized insects that use these chemicals as cues to feed or lay eggs? The answer intersects with my own scientific life.

When I accepted an offer for admission to the PhD program in biology at the University of Missouri–St. Louis, the fact that they'd awarded me the Peter Raven Fellowship in Tropical Biology was a big factor. My heart was set on studying insects in the tropics, and this was my golden ticket given UMSL's high-caliber graduate training program in tropical biology.

In 1964 the botanist Peter Raven, the namesake of that fellowship and longtime director of the Missouri Botanical Garden, and entomologist Paul Ehrlich published a research paper that built on Fraenkel's 1959 paper. In "Butterflies and Plants: A Study in Coevolution," Ehrlich and Raven proposed that when a plant evolves a new toxin, it will be protected from herbivores—for a while. During that phase, with its new chemical shield, the plant gets a leg up over competitor plants without the toxin, spreads across the landscape, and diversifies.

As its range expands, new species bud off from the old, protected by their new chemical defenses. In due course, insects catch up by evolving new resistance mechanisms that allow them to overcome the plant's defenses and to colonize the plants. Under this theory, insects specialize because there is no such thing as a free lunch—they either evolve adaptations needed to overcome novel plant toxins or they die out.

This is the first step of the cycle. Once insects have pierced the chemical shield, they will use the presence of the toxins as a chemical signpost for finding the "right" host plant. The toxin-resistant insects can then escape from their own competition and diversify into the new toxic niche. New plant species beget new insect species. Hence the term *coevolution*. The plants evolve because of the insects, and the insects evolve because of the plants, and so on. This process may explain why over half of the known species of life on earth are plants and the insects that eat them. May Berenbaum's studies on furanocoumarin-producing plants and their herbivores was one of the first real tests of the idea.

Raven and Ehrlich weren't the only biologists with a compelling explanation for the astonishing diversity of herbivorous insects with which we share a planet. Although they agreed that plant toxins and their precursors protect plants from attack by most herbivores or pathogens, ecologist

Catherine Graham and entomologist Elizabeth Bernays proposed in 1988 that it was enemies in the skies above and soil below that pushed herbivores to specialize, each on a different set of toxic host plants.

Graham and Bernays's idea was that after an insect colonizes a new toxic plant species and specializes, it flies under the radar of the old enemies, cloaked by new chemicals in an "enemy-free space." After a while, many specialist herbivores could even begin to sequester high levels of toxins from their host plants to defend against enemies.

In the end, both ideas, coevolution and enemy-free space, and some even newer ones, help explain why there are so many toxic plants and toxin-specialized herbivores. Each hypothesis relies on Fraenkel's idea that seems so obvious in hindsight: some chemicals made by plants are now used mostly for plant defense. These defensive chemicals are eventually overcome and then co-opted by specialized insects as weapons and defenses of their own. This chemical war of nature, which has raged for hundreds of millions of years, has also given rise to much of the pharmacopoeia that we use and abuse.

Now that we've established how the smell of death produces the same chemicals that are starting points for the alkaloids, and why nature's toxins are so diverse, we can now dive into how alkaloids specifically evolved to be so diverse and why they influence our own lives more than any other of nature's toxins.

Dank Chimneys and Dieffenbachia

As we have now learned, the chemical language used by plants and fungi to communicate with animals is ripe for mischief. This mischief extends to plant-pollinator interactions, too.

There isn't always an "orgy of mutual benefaction" between plants and their pollinators, as the late ecologist Sir Robert May once quipped. Plants, which can be robbed of their nectar by animals that don't move pollen on their behalf, can turn the tables and cheat animals out of an undeserved reward.

Plants lose pollen because pollinators eat some of it. It's hard to blame the pollinators. I've tried pollen collected by the honeybees I keep in my backyard hive. It tastes just like Nerds candy—this is less surprising than you might think, as malic acid is the main ingredient in both pollen and Nerds.

Pollinators also cheat by cutting holes at the base of flowers and stealing the nectar that way. They go to the trouble because this cheating behavior requires less energy of the animal than is required to enter a flower properly, getting pollen rubbed on them in the process.

As we have established, plants are not innocent bystanders when it comes to pollination. Manipulation of the animal mind and body is the name of the game for plants that need to attract pollinators, whether through reward or deception—and sometimes plants are the ones doing the cheating.

Small arums like skunk cabbage and the lords-and-ladies plant live in the temperate climates of the Northern Hemisphere and are encountered by millions of people each spring. Both plants synthesize chemicals that mimic the smell of death. In fact, the lords-and-ladies arum emits cadaverine and putrescine into the air. There is a remarkable connection to alkaloids here. Alkaloids are made from cadaverine and putrescine, as are many of our neurotransmitters, including those like serotonin, dopamine, and norepinephrine.

In Sumatra, the titan arum is known by its Indonesian name *bunga bangkai,* or "corpse flower." It too smells like rotting flesh. What's more, the flower's ten-foot-tall spike heats up to around the temperature of the human body the night after the flower opens. The titan arum produces the largest flowers of any plant in the world. Waves of heat propel the scent up its fetid chimney and then throughout the rain forest.

The putrid scent and heat of the corpse flower, the skunk cabbage, and the lords-and-ladies arum evolved to attract carrion-feeding insects in search of rotting animal tissues in which to lay their eggs. The animals are fooled into moving pollen between the flowers. This is called a *deceptive pollination system* because the signal produced by the plant evolves to mislead its target. Arums don't reward the insects that move pollen on their behalf; the plants make a false promise. Only the plants benefit from the interaction. This situation is more like competitive, predator-prey or host-parasite dynamics in which there is a clear winner and loser.

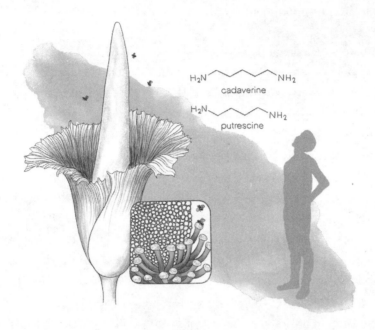

The skunk cabbage, lords-and-ladies arum, and corpse flower are not the only arums that play this kind of game with insects. The arum family is replete with species that produce all sorts of scents designed by evolution to trick animals, from bats to beetles, into visiting their flowers. Bouquets of anise, banana, booze, burned rubber, cheese, chocolate, dung, petroleum, polish, and even fried fish are released by arums.

Neuroethologist Marcus Stensmyr and collaborators discovered that the same fruit flies that I and many other biologists use as model organisms are naturally lured into entering Solomon's lily (in Arabic, called *mekhalet el-ghoule*, "paintbrush of the witch"), which has labyrinth-like flowers that look like black calla lilies. On the day the flower opens, flies that enter become trapped at the base of the spadix, near the pistils. (The pistils are the female parts of the flower and are receptive to pollen a fly carries in from other Solomon's lily plants it might have visited before.) The next day, the anthers, the male reproductive parts just above the base of the spadix, produce pollen, which gets sprinkled onto the captive flies. The anthers quickly wilt, allowing the pollen-covered flies to leave and—hopefully, from the plant's point of view—find a newly opened lily that the flies will be fooled into entering and pollinating.

You may be wondering why the fruit flies arrive at the flowers in the first place. You already know the answer: they are called fruit flies for a reason. The Solomon's lily deceives fruit flies by producing odors that mimic those emitted by the fermenting yeast and ripe fruit on which they normally lay eggs. This scent targets a set of odorant receptors in the fly. The odors that these receptors are tuned to will normally attract egg-laying females to the food required by their larvae.

Children can also be deceived by the alluring Solomon's lily. Encounters with this plant drive more calls to the Israeli poison control center than do any other.

Like most arums, the tissues of Solomon's lily are protected against herbivores by poisonous oxalate crystals, which are potent kidney toxins. To make matters worse, the crystals are packaged up in the tissues of the leaves, stems, and flowers as raphides, which look and function just like microscopic needles. As a child begins to chew on any part of the plant, the cells lining the mouth and tongue are punctured by the raphides. The result is numbness, drooling, and inability to speak. If enough tissue is ingested, gastric distress and even airway closure can occur. Many arums, including species of *Philodendron* and *Dieffenbachia,* are common houseplants. Because they can also cause distress if their leaves are chewed, they should be placed away from young children or inquisitive pets.

Although not an alkaloid, oxalate can be linked to the production of alkaloids. It is synthesized by plants from oxaloacetic acid, which the plant in turn uses to produce the amino acid lysine. Plants then use lysine to make the alkaloid piperidine, which lies at the molecular heart of many alkaloids that are important to us, from piperine in black pepper to morphine in the opium poppy.

The smell of death is produced by bacteria that break down amino acids into molecules like cadaverine, putrescine, and skatole. Most organisms also synthesize these chemicals to make alkaloids, whether for use as poisons, signaling molecules, or neurotransmitters. For example, as discussed, the cadaverine and putrescine that some plants synthesize help lure insects toward the plant and consequently help disperse pollen. As products of decomposition or naturally synthesized pollinator attractants, these alkaloid precursors play two-sided roles in the war of nature.

Now that we've looked at the role of alkaloid precursors, let's dive in and explore humans' relationship with alkaloids. Amphetamines, ayahuasca, Ecstasy, nutmeg, magic mushrooms, and toad venom alkaloids are next.

Myristicin, Ma Huang, and Methamphetamine

As we've just seen with arums, chemical signals produced by plants can prevent or promote visits from animals. This pattern repeats itself across the layers of tissue in fruits that, like the papaya, are laden with chemical reward and punishment.

The skin of most fruits is thick and protective. The animals that can figure out a way to get in are often rewarded by nutritious flesh just underneath. Beyond the skin and flesh lies another well-defended layer studded with toxic seeds that protect the embryos within. There are good reasons why you spit seeds out before chewing further.

If broken open by an unlucky bite, papaya seeds produce burning mustard oils just like those in wasabi. Similar strategies are used by most of the plants whose seeds we use as spices, including allspice, anise, caraway, cardamom, coriander, cumin, fennel, peppercorns (black, green, Szechuan, sansho, and pink), mustard seeds, nutmeg, red pepper flakes, turmeric, wasabi, and many additional ingredients that go into curry powders.

The chemicals we seek out in spices didn't evolve to enliven our food and drink—they evolved because they benefit the plant. In the tiny doses most of us use them, they are harmless. But in larger doses, and to smaller animals, spices contain chemicals that are both distasteful and toxic. Many of these chemicals are alkaloids. So instead of crushing the seeds and eating those, animals that eat fruit, if they are able, spit them out. If not, the seeds are often regurgitated or excreted in the scat. Depending on the gut passage time, the seeds can unwittingly be dispersed afar. Birds, mammals, reptiles, and even fish, especially those that take advantage of seasonal flooding in the Amazon, are among the most important seed dispersers.

From the plant's perspective, then, enticing frugivores is worth the energy needed to produce the reward. This is another example of a win-win kind of coevolution, much like that between orchids and orchid bees. Instead of moving pollen, frugivores move seeds in the exchange of goods and services, as in the tripartite interactions between fruit, yeast, and primates.

Just how advantageous it can be for a seed to end up in the gut of an animal is remarkable. As a UMSL graduate student studying in Ecuador, ornithologist Kimberly Holbrook showed that when toucans eat a seed from the beautiful South American nutmeg tree *Virola flexuosa*, they swallow the seed whole, digesting only the bright red fatty aril surrounding the seed. Had the toucan crushed the seed while eating the aril, the bird would have received a dose of poison.

The toucan may hold the seed in its body for well over an hour. Holbrook showed that the large toucans can fly more than a mile in that time. From the seed's perspective, this aerial transport is a good thing; were it to sprout near its parental tree, it would be attacked by the same specialized microbial pathogens and insect herbivores.

Life in the tropical rain forest is difficult for plants, given the hordes ready to attack. So, plants fight back using toxins, and we are the beneficiaries.

Some nutmeg seeds, like the nutmegs we use as spices from the Maluku Islands (called the Spice Islands by European colonizers) of Indonesia, contain the terpenoid myristicin, which protects the embryo of the nutmeg seed from attack. In the human body, myristicin may be converted into a psychedelic amphetamine called MMDA (3-methoxy-4,5-methylenedioxyamphetamine), which is sometimes confused with the club drug MDMA (3,4-methylenedioxymethamphetamine), also known as Ecstasy or Molly. MMDA and MDMA have similar chemical structures but exert different effects on the body.

Although myristicin isn't an alkaloid, we can lump amphetamines like MMDA and MDMA with alkaloids.

The physician Andrew Weil wrote his undergraduate thesis at Harvard on nutmeg as a narcotic. He concluded that although a person could get high from nutmeg, possibly owing to the MMDA transformed by the body after ingestion of myristicin, so much of the spice must be consumed to create a narcotic effect that doing so would make the person sick. However, in smaller doses, nutmeg has been widely used traditionally for thousands of years in South Asia and Southeast Asia to treat all manner of illnesses, including diarrhea and insomnia.

Amphetamines generally work by inhibiting the natural recycling of neurotransmitters like dopamine and norepinephrine by neurons in the brain. This action leaves higher standing levels of neurotransmitters outside the brain's nerve cells and causes these cells to fire repeatedly. The overfiring drives higher brain activity, which can enhance some aspects of cognitive functioning. This increase in the availability of dopamine and norepinephrine leads to higher energy, focus, and self-confidence, as well as euphoria. But being in overdrive eventually comes at a cost. Drugs like methamphetamine are addictive because of their reinforcing effect on the brain's reward system due to the euphoria a user experiences.

Amphetamines are also made from the plant ma huang, known scientifically as *Ephedra sinica*, and its relatives, which have been used in East Asia and South Asia for five thousand years as medicinals. In 1885, Japanese chemist Nagayoshi Nagai was the first to isolate the alkaloid ephedrine from ma huang, followed by chemists at the German pharmaceutical giant Merck a few years later from a closely related species. Still later, ephedrine became the first asthma medication because it relaxes the muscles controlling the bronchial tubes of the lungs. Ephedrine is a stereoisomer of pseudoephedrine—the active ingredient in some of the most widely used cold and cough medicines.

Ephedrine can easily be turned into methamphetamine (meth) with just a few chemicals. Therefore, in 2005 the US government passed the Combat Methamphetamine Epidemic Act, which required pseudoephedrine products to be sold behind the counter, with strict limits on the number of pills a person could purchase.

In 2020, some 2.5 million people in the United States aged twelve and above said that they had used meth that year. This figure only applies to meth and not the prescription-based amphetamines. It is a sobering statistic.

Even more troubling is that over half of the adults in this group had methamphetamine use disorder. In other words, they could not easily stop using the drug even if they wanted to. Increasingly, meth is being combined with opioids like fentanyl for street use. The combination of meth with this exceedingly strong opioid has contributed to increased fatal overdoses from 2014 to 2020 and beyond.

Like ma huang, a plant called khat (*Catha edulis*), in the bittersweet family Celastraceae, has been used in East Africa and the Arabian Peninsula for thousands of years as a stimulant and anti-fatigue aid. Tens of millions of people, mostly men, still chew khat daily in these regions of the world. Khat contains the amphetamine-like alkaloid cathinone or beta-keto-amphetamine. Chemically modified cathinones include the antidepressant prescription drug bupropion (Wellbutrin). Other chemically modified cathinones are used to make the street drug known as Bath Salts in the United States.

Cathinones work on the brain in much the same way as meth and other amphetamines. They prevent the natural recycling of some neurotransmitters like dopamine, norepinephrine, and serotonin in the brain and even increase their production.

Although the alkaloids described in this section alter the mind and consequently are psychoactive, they tend to exhibit no or mild psychedelic effects, with the exception of MMDA. Another set of alkaloids, on the other hand, famously induces hallucinations and are known as psychedelics. These alkaloids have been used by many Indigenous cultures around the world as entheogens, that is, psychotropic drugs used for spiritual practice. However, the line between medicinal and spiritual is blurred in many of the cultures that regularly use these substances.

DMT, *Bufotenine, Psilocybin, and Mescaline*

The tryptamine and phenethylamine alkaloids include some of the most intriguing, taboo, and en vogue of mind-altering drugs. They are also some of the most poorly understood by science.

We will begin with the tryptamines. One such alkaloid is found in *Virola theiodora*, an Amazonian nutmeg related to the one Holbrook studied in Ecuador. The resin in its bark is the source of a mind-altering snuff and arrow poison used by the Yanomami people in Venezuela and Brazil. In that resin is one of the most potent psychedelics known: *N,N*-dimethyltryptamine, or DMT.

DMT and other related tryptamines are found in at least twelve plant families and in animals and fungi. The tryptamines bufotenine and O-methyl bufotenine are made by the skin of some toads as well as by several plant species. Magic mushrooms contain the tryptamine psilocybin and its psychoactive derivative psilocin. I will refer to these diverse tryptamines as DMTs unless noted because each has a DMT molecule at its core.

Clearly, DMTs are not the *only* potential toxins found in the skin of the toads that make them. Although consumption of these tryptamines is rarely deadly, cardiac glycosides like marinobufagenin are found in toad glands, too, and if ingested at the right concentration, they can cause illness and even death.

The danger of using extracts of toad glands to trip might be why, despite the lore associated with toad venom, there is scant historical evidence of Indigenous peoples of the Americas using it. However, some Indigenous communities there have now incorporated toad venom into their practices, particularly as a treatment for methamphetamine and other drug use disorders.

In fact, the practice of smoking toad venom was unheard-of until 1983, when Ken Nelson, then an artist from Denton, Texas, milked the skin of a Sonoran Desert toad because he had read that chemists discovered bufotenine in that species. Nelson then wrote a pamphlet titled "*Bufo alvaris:* The

Psychedelic Toad of the Sonoran Desert" under the pseudonym Albert Most.

While there are paintings in ancient Mayan structures depicting toads, little else links the modern practice with an older one, although more ancient use cannot be ruled out. On the other hand, abundant evidence shows us that the ancient peoples of the Americas sourced DMT from particular plants and fungi that, unlike toads, contain no highly poisonous cardiac glycosides.

DMTs bind to our brain's serotonin receptors, or 5-HT receptors, so called because serotonin is an endogenous tryptamine alkaloid, 5-hydroxytryptamine, or 5-HT. They have an affinity for a specific type of 5-HT receptor called 5-HT$_{2A}$, which I'll later show may be related to alcohol cravings.

DMTs are yet another example of how different human cultures around the world independently evolved similar uses for the same chemical defenses produced by other organisms. The chemicals become available for such practices because they evolved independently in many species in different places. A few examples showcase the nexus of these two phenomena.

The velvet bean, a vining legume shrub from equatorial Africa, produces a bevy of tryptamine alkaloids, including serotonin, bufotenine, and DMT in trace amounts and much larger quantities of L-DOPA, a precursor to catecholamine neurotransmitters like dopamine. Velvet bean is used by various Indigenous cultures in Africa as a medicinal and an arrow poison.

Across the Atlantic, in Latin America and the Caribbean, bufotenine is found in two other legume tree species in the genus *Anadenanthera*: *A. vilca* and *A. yopo*. These plants have been used for millennia by Indigenous peoples of the Americas as psychoactive snuffs and as an ingredient in some types of chicha, a maize-based and widely consumed drink of ancient Andean origins.

Bufotenine, probably derived from an *Anadenanthera*, was found in the hair of two 1,500-year-old mummies from Chile. More recently, residues of bufotenine, DMT, and psilocin, along with the alkaloids cocaine and harmine, were scraped from the inside of a ritual pouch found in south-western Bolivia in 2008 and made from the snouts of three foxes around 1000 CE. This archaeological evidence shows that humans have been using DMTs for millennia.

DMT is also the principal psychedelic in ayahuasca—a Quechua term for "vine of the gods." This chemical is used as a medicinal and in shamanistic rituals of a diverse group of Indigenous cultures of the Amazon and the Andes. Ayahuasca is prepared in countless ways, but the principal ingredients often include tissues from two sets of plants. One set of plants produces DMTs and includes species of *Psychotria* from the coffee family or *Diplopterys* from the Malpighiaceae family. A different set of plants produces alkaloids called beta-carbolines, or harmine alkaloids, and includes plants in the genus *Banisteriopsis,* which also belong to the Malpighiaceae.

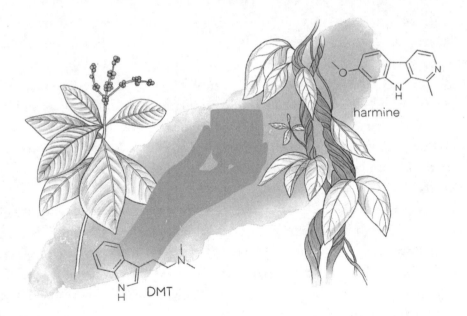

harmine

DMT

If plants from each type are not combined in the concoction, there will be no psychedelic effect. The reason is that our body's monoamine oxidase (MAO) enzymes quickly deactivate DMT when the tryptamine is taken as a drink. To get around this firewall, South American shamans ingeniously use the fact that beta-carbolines *inhibit* our MAO enzymes, thereby preventing MAO from deactivating the DMT. This MAO activity inhibition is where the *I* in MAOI comes from.

When beta-carbolines are included in the mixture, the DMT is then free to travel through the gut and into the blood, where it eventually makes its way to the brain, binds to the serotonin receptors, and works its magic. In contrast, when toad venom is smoked, its DMTs slip past the MAOs in the gut because it reaches the brain through the lungs.

Psilocin is ultimately detoxified by the MAOs, but unlike most DMTs, it is just naturally more resistant to the MAOs. *Psilocybe* mushrooms, also called magic mushrooms, have another trick up their sleeves. They produce both the psilocybin *and* the beta-carbolines — the psychedelic substance and the MAOIs. So, when animals consume these mushrooms, the psilocybin cannot easily be deactivated by their body's MAOs, thanks to the beta-carbolines that are also in the mix. This biochemical interaction is an example of biological evolution and human cultural evolution both creating the same innovation — concoctions of both DMTs and MAOIs.

Nowhere were *Psilocybe* and the many other psilocybin-producing mushrooms more important in human culture than among the Aztecs. Accordingly, these fungi were called teonanacatl, or "flesh of the gods," and were consumed in a variety of healing and religious ceremonies. Their traditional use continues in several communities throughout Mexico.

Two other important serotonin receptor-binding psychedelics deserve mention. The first are the ergot alkaloids. Ergot is both the name of a plant disease caused by fungi of the genus *Claviceps* and the term for the club-shaped fruiting structure of these fungi. Ergot alkaloids have long been used to treat migraine headaches and postpartum bleeding. These chemicals are derived from lysergic acid and include the synthetic relative lysergic acid diethylamide (LSD), which will be further discussed later in the book.

When grains like rye are contaminated with ergot fungus and are eaten, especially chronically, a poisoning disease called Saint Anthony's fire can develop. The disease is caused by the constriction of the blood

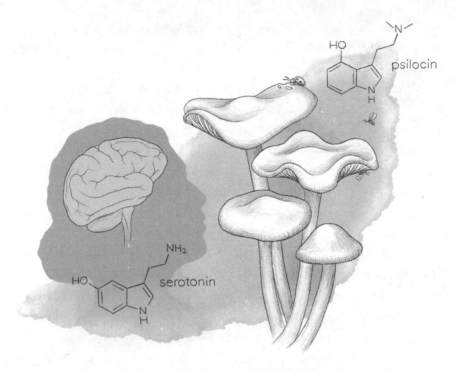

vessels. Ironically, then, a medication that is a cure for postpartum bleeding can also cause a deadly disease in another situation.

Another important psychedelic alkaloid that binds to the serotonin receptors is mescaline. This alkaloid has more in common chemically with ephedrine and amphetamines but shares more functionally with the DMTs and ergot alkaloids.

Mescaline is found only in the Americas in two main types of cacti: the smaller peyote cactus (*Lophophora williamsii*) and the towering, columnar San Pedro cactus and other cacti in the genus *Trichocereus*. While peyote is used by the Indigenous peoples of Mexico, the San Pedro is used by the Indigenous peoples of South America. Later, peyote's use spread to the Indigenous peoples of the United States and Canada as well. It has been used for at least fifty-seven hundred years for a variety of purposes, including ritual healings, as I will discuss later when we examine why humans may have initially used these psychedelic drugs in general.

Most of the DMTs, natural and synthetic ergot alkaloids, and mescaline are Schedule I drugs in the United States. This classification means that under federal law and many state laws, these drugs or the organisms that make them cannot be legally purchased, possessed, or distributed. The same classification and laws hold in many other countries. Nonetheless, these alkaloids are having a moment as psychoactive drugs of choice for recreational, medicinal, and spiritual use.

Why psychedelic alkaloids evolved in the first place is, of course, a central question. You are probably thinking that they may serve as defensive chemicals to protect against attack. The truth is that we really don't know.

Among the plants that make DMT, some evidence from reed canary grass and its relatives in the genus *Phalaris* suggests that the chemicals protect against grazing mammalian herbivores like cows, sheep, and even kangaroos, which find the alkaloids relatively unpalatable. The "phalaris staggers," reported from animals that feed on these plants, reflect a psychotropic effect. An increase in diarrhea is also observed.

For magic mushrooms and psilocybin, one clue to how they evolved to produce psychedelics comes from their blue color. Magic mushrooms turn blue when wounded, and psilocybin is responsible for this change of color. When the mushroom is wounded, two enzymes chemically transform psilocybin into a chain of psilocin molecules that become bonded to one another. When formed, these chains of psilocin molecules act much like tannins, which also turn blue when oxidized. These blue chemicals probably produce gut-damaging free-radical oxygen molecules when the insects eat the fungi.

So, like the chemicals that gave us iron oak ink and purple ice, the magic mushroom's tannin-like chemicals may have a similarly toxic effect on fungus-feeding insects. In this context, psilocybin is a prodrug like cyanogenic glucosides. Both psilocybin and cyanogenic glucosides are nontoxic in plants but become transformed into a toxin like cyanide once wounding occurs. Because the chains of bonded psilocin molecules are probably also toxic to the mushrooms, there is an evolutionary advantage to keeping psilocybin in its prodrug state, with the fuse unlit, so that the chemicals don't harm the mushroom.

Although this defense theory of why magic mushrooms make psilocybin has a lot of merit, psilocin may also have psychoactive effects in the insect brain. We know this because cicadas infected with a fungus that appears to produce psilocybin forgo their own reproduction and become oversexed zombies even after their genitalia fall off, owing to the fungal infection. This pseudo mating behavior spreads the fungal spores to other cicadas and benefits the psilocybin-producing fungus.

With this observation, we can see how mushrooms that produce psilocybin might make the chemical to manipulate insects into transporting their spores, much in the way that flowering plants use mind-altering chemicals to coax pollinators to move their pollen. It remains to be seen which, if any, of these hypotheses—the anti-fungivore theory or the spore-dispersing one—accurately explains the evolution of DMTs like psilocybin.

Some scientists have proposed that because ergot-infected grasses are quite colorful, the fungi may be warning animals to stay away. In other words, ergot might have evolved a similar overall strategy like that of the monarch butterfly, whose bright colors warn predators of the toxins within.

A similar question exists for mescaline. Perhaps plants that evolved to synthesize this psychedelic compound are better at dissuading herbivores or attracting pollinators. But we still don't know.

Given that DMTs, ergot alkaloids, and mescaline are bitter, it is likely that most animals avoid them in nature after taste-testing the plants, fungi, or animals that make these substances. So the question is, why do humans, in culture after culture, use serotonin-receptor-binding drugs like these? We will return to this important question a bit later. But for now, we can already see that shamanistic ritual healing is certainly one explanation. However, these chemicals arc also used recreationally, at least nowadays. The visual hallucinations they trigger in the brain may themselves be the reward.

Corroborating this idea are experiments with rhesus monkeys. Animals that were housed in a dark box with self-administered lettuce cigarettes (originally designed as a nicotine-free cigarette for people wanting to quit smoking) containing DMT. If the animals were not held in complete darkness, they avoided the DMT cigarettes. The scientists who ran these experiments think that the DMT caused the light-deprived monkeys to experience an "internal hallucinatory light," which itself became a reward.

There is no evidence that animals in nature use DMTs or mescaline for their own benefit. Nor do laboratory animals. Except in very specific situations, laboratory animals find these chemicals repulsive, whether they are in food, drink, or inhaled smoke.

Even more astounding than monkeys smoking DMT-laced lettuce cigarettes, brains of mammals, including rats and humans, produce DMT, too, in small amounts. I've already discussed how the human body makes salicylic acid and cardiac glycosides or molecules that mimic their effects. So, maybe the presence of endogenous DMT in our brains isn't much of a surprise. Still, the lesson is remarkable: plants, fungi, and bacteria have repeatedly evolved to deploy molecular mimics of our neurotransmitters and hormones or to produce chemicals that disrupt their production and movement in our bodies.

Alkaloids like the DMTs and mescaline are typically not drugs of abuse. Yet, their use is generally taboo and highly restricted in most of the world. This proscription may be changing in light of new information on their role as potential cures. The curative role is one in which these chemicals have been used for millennia.

Returning to Our Roots

Abstinence and long-term sobriety are difficult to achieve for many with drug use disorders. Drugs developed by Big Pharma to treat these disorders, in combination with group and talk therapy, provide some hope. However, natural and some synthetic psychedelics are among the most promising new treatments for drug use disorders and often co-occurring mental health disorders, including severe attention deficit hyperactivity disorder, obsessive compulsive disorder, PTSD, depression, and anxiety.

Given the promise of psychedelics, I can't help but think that maybe my father didn't have to live with AUD his whole adult life and die from its complications. It pains me that these promising advances in the treatment of AUD didn't appear until after his death. But it also gives me hope for others who might be dealing with drug use disorders and other treatment-resistant mental disorders but who are still alive.

The drugs that may be our best hope for treating these disorders are serotonergic psychedelics. They include the tryptamines DMT; psilocybin; lysergic acid derivatives like LSD; and substituted phenethylamines like mescaline, MDMA, and analogs. Indigenous shamans have relied on serotonergic psychedelics for millennia to treat illness, whether it is physical, mental, or spiritual. These drugs all tap into the 5-HT_{2A} serotonin receptors in the brain.

Because of the great hype surrounding these chemicals, we need to scrutinize the hard evidence, of which there is precious little. However, the results of preliminary clinical trials, the gold standard in medicine, are compelling, despite the small sample sizes.

A small placebo-controlled trial using ayahuasca (containing DMT) to treat treatment-resistant depression was reported in 2018. The participants receiving the ayahuasca reported significantly reduced depression severity within one day of the treatment compared with people taking the placebo. The effect persisted through the one-week study period.

LSD showed great promise in treating AUD in the mid-twentieth century, but backlash against the counterculture that used it for other purposes snuffed out progress by the 1980s. However, between 1966 and 1970, six randomized, controlled studies using LSD were conducted on people with AUD. Early on, after one to three months, there was near-complete abstinence from alcohol use among those who received the single dose of LSD in the trial. After three to six months of the participants' receiving a single dose of LSD, there was a twofold reduction in risk of alcohol misuse.

As a natural product from magic mushrooms, psilocybin, although a Schedule I drug in the United States, may carry less social stigma than LSD does, and it may be even more promising as a treatment for some drug use disorders. Up until 2022, peer-reviewed, published studies that used psilocybin to treat tobacco addiction and AUD have been open label (which means patients know they are getting it and not a placebo), with small sample sizes. In both cases, the results were promising.

Then in September 2022, the first randomized, double-blind placebo-controlled study was published. It used diphenhydramine as a placebo (forty-six patients) and psilocybin (forty-nine patients) as a treatment for AUD, in combination with talk therapy for both arms of the study. Two doses of the drug or placebo were given, one at four weeks and another at eight weeks, and the study ran for thirty-two weeks. The results were clear: the percentage of heavy drinking days was approximately 10 percent in those receiving psilocybin compared with around 24 percent for the placebo arm. The average number of drinks per day was also lower in those receiving psilocybin than in the placebo group. So, psilocybin seems to work at mitigating AUD in people who are highly motivated to receive treatment, at least. The question is how. The answer might be found in the brain circuits involved in craving.

During the year my father abstained from drinking, my mother noticed that as he sat in his chair in the living room, he physically struggled to refrain from going to the refrigerator to search for alcohol. He was miserable, though, and slowly began to incorporate alcohol back into his daily rhythm until he was right back where he started.

Many parts of the brain are involved in perpetuating AUD and drug use disorders. Cravings are a central part of why these illnesses are so difficult to treat. Crucially, we now know that for many if not most people with AUD, chronic alcohol use reduces the number of glutamate receptors in the prefrontal cortex of the brain, which controls executive functioning (like deciding to go to the fridge to get a beer or to stay put in your chair).

These glutamate receptors are located within neurons that project through the brain to the dopaminergic mesolimbic reward system, which mediates the anticipation and the reward associated with the drug. Remarkably, the 5-HT_{2A} receptors in the prefrontal cortex pair up with the glutamate receptors in these neurons. When drugs like psilocin bind to the 5-HT_{2A} receptors, they cause the glutamate receptors to activate. So, taking psilocybin, which is transformed into psilocin in the body, reactivates a crucial governor that controls the alcohol cravings—at least in laboratory mouse models of AUD. Whether this mechanism explains why psilocybin causes less drinking in people with AUD is unclear.

Ironically, an addiction to one natural toxin may be ameliorated by another. On the other hand, this apparent paradox is just another example of the duality of many of these chemicals in our hands. The poison can be the cure. Although there is no guarantee, the life-giving and life-ending powers of these chemicals can be seen as two sides of the same coin.

An important caveat is that treatment of drug use disorders and other mental illnesses is serious business that should not be attempted without clinical supervision. These drugs are not panaceas; they are tools that modern medicine is just starting to learn how to deploy, often building on thousands of years of Indigenous practice.

For some people, especially those with schizophrenia, their illnesses can also be made worse by some of these psychedelics. Furthermore, the plants, mushrooms, and animals that produce these drugs, along with the pure forms of the drugs themselves, fall under the legal definitions of controlled substances. As such, their possession and use may be illegal, depending on the jurisdiction in which you live.

As I've pointed out, none of nature's toxins evolved for our sake. They were here long before us and in many cases keep enemies at bay — they are toxins first, *potential* drugs second. As Shawi elder and traditional healer Rafael Chanchari Pizuri of the Peruvian Amazon explained: "The power of these plants can cause human losses."

You may not think of them this way, but the most widely used and socially accepted psychoactive alkaloids are caffeine and related methylxanthines, which are found in coffee, tea, chocolate, kola nut, yerba maté, and several other distantly related plant species. Another psychoactive alkaloid that remains widely used and socially accepted is nicotine, found in tobacco and other plants in the nightshade family. So important are these two alkaloids that I've devoted a whole chapter to them. We'll examine caffeine and nicotine in the next chapter.

7.

Caffeine and Nicotine

No suitor comes in my house
unless he has promised to me himself
and has it also inserted into the marriage contract
that I shall be permitted
to brew coffee whenever I want.

— JOHANN SEBASTIAN BACH, *COFFEE CANTATA*

Filter It

Just before bed, I precisely measure out the amount of coffee Shane and I will need to make it through the next day as I ready the automatic drip machine. If I prepare too little, I will be sleepy and unable to mentally focus. If I overshoot, I will be anxious and feel sick.

At lower doses, caffeine causes euphoria and increases our alertness and cognitive performance. At higher doses, it causes nausea and increases anxiety and overall shakiness. I am addicted to caffeine and couldn't feel better about it. You'll see why.

Although I used to use a French press, I now *only* make coffee with an automatic drip machine or by pour-over. I made the switch to making only filtered coffee after reading a 2020 study of more than half a million people in Norway that found the adults who consumed unfiltered coffee had a significantly higher chance of dying over twenty years than did those who drank filtered coffee or no coffee. The increased mortality risk is probably associated with drinking unfiltered coffee, at least partly due to the presence of two coffee terpenoids that are largely removed during filtration. Both are associated with increased cardiovascular and heart disease because they raise cholesterol levels.

The offending terpenoids, cafestol and kahweol, elevate low-density lipo-protein (LDL, or "bad") cholesterol levels in our blood. In fact, cafestol is the most potent LDL-inducing chemical known from the human diet. The mech-anism, at least according to studies in laboratory mice, is the binding of these two sterols to a hormone receptor in the small intestine. This binding sends the wrong message to the liver, which then cranks up the production of LDL.

Scandinavian boiled coffee, French press (cafetière) coffee, and Turkish coffee have the highest levels of these terpenoids. Espresso and mocha have more modest levels, and instant, percolated (with a filter), and filtered coffee have the lowest, by a wide margin: unfiltered coffee contains thirty times more cafestol and kahweol than does filtered coffee.

I found several articles on coffee or wellness-related websites claiming that reusable metal mesh filters (often stainless steel or 24k plated gold) are less effective at trapping kahweol and cafestol than paper filters are. I could find no evidence supporting this claim in the peer-reviewed literature. But absence of evidence is not evidence of absence.

So, I dug into this question of the effectiveness of metal filters. Instead of an absence of evidence, I found a rigorous experimental study that resolved the question of whether metal mesh filters did the trick at remov-ing these terpenoids. When used in automatic drip machines, paper and metal filters (the investigators used Swissgold brand of mesh filters) were equally effective in preventing most kahweol and cafestol from passing through to the brew below. However, there is a catch. In this study and a few others, pour-over coffee from a metal mesh filter produced higher levels of kahweol and cafestol than did coffee from an automatic drip machine using the same kind of filter. Although speculative, the cause of this differ-ence may be that the water added to the automatic drip machines drips slowly and gently onto the ground coffee in the filter below, allowing a thick "filter cake" to form on the bottom, potentially acting as a kind of first-pass filter itself. During pour-overs, on the other hand, a stream of water rapidly causes the grounds to continue to swirl and remain suspended, likely allowing more of the smaller-sized particles rich in the terpenoids to escape through the mesh. Why didn't we get to do experiments like this in home economics class?

I know what you are going to ask next: what about coffee pods? Pods have a paper filter built into them, so the cafestol and kahweol levels are comparable to automatic drip coffee prepared with paper or metal mesh filters or pour-over coffee prepared with a paper filter. What's more, the characteristics of the bean (variety, roast), the water temperature, the grind size, and the water-to-coffee ratio all affect the levels of terpenoid in coffee.

I was convinced enough by what I gleaned from the scientific literature to change my habits. I stopped my twenty-year practice of using a French press to make coffee. Although I occasionally splurge on an espresso drink (which has only modest levels of the terpenoids) when I'm out and about or traveling, at home I now only make filtered coffee using the automatic drip machine with a gold mesh filter. Or I make pour-over coffee using a paper filter—and an unbleached one at that.

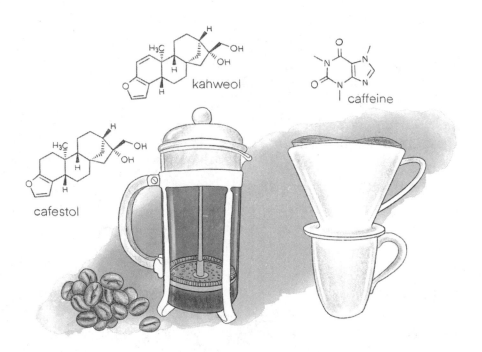

kahweol

caffeine

cafestol

In the United Kingdom, people generally drink coffee that contains low to modest levels of cafestol and kahweol (filtered coffee, instant coffee, or espresso). A UK study followed 171,616 people from 2009 to 2018 to determine whether drinking coffee was associated with reduced risk of death, and if so, by how much. In general, those who drank coffee had a lower risk of death than those who did not, and this finding held for the risk of dying from cancer or cardiovascular disease.

But the devil is in the details. A slightly reduced risk of dying was found for those who drank up to 2.5 drinks per day compared to those who didn't drink coffee. The people drinking 2.5 to 4.5 coffee drinks per day were 29 percent less likely to die during the study than were people who drank no coffee. Above 4.5 drinks per day, and the risk of dying during the study was the same as those who consumed up to 2.5 drinks per day.

Those in the middle of the spectrum (i.e., those who drink 2.5 to 4.5 cups a day) were the least likely of all the participants to die. This study comports with a much larger umbrella review of studies that found that drinking 3 to 4 cups per day is associated with a roughly 17 percent lower risk of death than the risk for nondrinkers.

Finally, there was a big potential downside to high caffeine or coffee consumption in pregnancy. Compared with low consumption of coffee, high consumption was associated, on average, with a 31 percent higher risk of low birth weight, 46 percent higher risk of pregnancy loss, and 22 percent higher risk of first trimester preterm birth, and 12 percent higher risk of second trimester preterm birth. There were no elevated risks observed in the third trimester. Chapter 10 will examine what may cause these patterns.

The protective association of coffee drinking is also observed for those who drink decaffeinated coffee. One explanation is that for most of us, coffee, whether caffeinated or decaffeinated, is the single largest source of antioxidants in our diet in the form of polyphenols. As discussed in an earlier chapter, the term *antioxidant* connotes chemicals that protect against oxidants, cellular stressors produced by our own bodies and the environment.

The potential protective effect of coffee in our diet may be largely due to the presence of these antioxidants in coffee beans. Among these, the chlorogenic acids are the prime suspects. Double-blind, placebo-controlled studies have found that chlorogenic acids in the diet can improve cardio-vascular functioning, reduce blood pressure, and decrease the risk of meta-bolic syndrome. Coffee drinkers consume up to 1 gram of chlorogenic acids per day; by comparison, a typical adult aspirin tablet weighs 325 mil-ligrams, or one-third the amount consumed by coffee drinkers.

This discussion of coffee brings us back full circle to earlier discussions of chlorogenic acids. These chemicals, the same phenolics associated with protecting our health when we drink coffee, are among the toxins we've seen in earlier chapters. Chlorogenic acids that end up in the fog drip of eucalyptus trees kill other plants growing near them. And these same chemicals, found in the salt grass that montane voles eat, cause the rodents to stop reproducing.

In the Norwegian study on unfiltered coffee, the higher mortality rates among unfiltered-coffee drinkers (compared with the filtered-coffee drinkers and nondrinkers) may be caused by the LDL-raising effects of the terpenoids cafestol and kahweol. On the other hand, the lower death rates among filtered-coffee drinkers compared with the nondrinkers may be caused by the polyphenols, including chlorogenic acids. So, coffee contains toxins that are associated with both positive and negative effects on human health.

A cautionary note is called for. In any observational study, whether the Norwegian one, the UK one, or the umbrella one, selection bias could give rise to the observed effects between coffee drinking and health risks. If selection bias came into play, then some other variable besides coffee drink-ing underlies the pattern because it correlates with people who consume different levels, and types, of coffee. Yet in light of animal studies, including some human research, the potential biological mechanisms for both the LDL-increasing aspects of the terpenoids and the cardiovascular and dia-betes risk-reducing aspects of the phenolics in coffee are clear. Still, we need to tease out all the factors by conducting a long-term, double-blind, placebo-controlled study, which is difficult to do.

Although I'm neither a nutritionist nor a physician and so do not give dietary or medical advice, I'm convinced enough of the totality of this information to change my own habits. Therefore, I will both filter it and continue to drink a lot of caffeinated coffee, although not *too* much. I drink three or four large cups of filtered coffee per day, with a cappuccino or flat white sneaked in there a few times a week as a treat.

Beyond the potential cardiovascular protection of coffee, coffee drinking is also associated with dramatically lower risk of Parkinson's disease in observational studies. Similarly, a large observational study in the United Kingdom found that drinking three to six cups of coffee or tea daily reduced the risk of dementia by 28 percent and stroke-induced dementia by 48 percent. What is driving these associations is unclear. One candidate is a serotonin derivative called eicosanoyl-5-hydroxytryptamide from coffee beans. The chemical can slow Alzheimer's disease progression in laboratory rat models of the disease. These findings certainly warrant further study.

With coffee's terpenoids and phenolics out of the way, we can finally get down to the big reason we consume caffeinated beverages like coffee: the alkaloid caffeine. We will now explore its origins, its biological effects, and our relationship with this alkaloid.

Brazen Beetles and Killer Coffee

Caffeine and the human mind seem like a match made in heaven. But even though billions of people imbibe caffeinated beverages each day, caffeine first evolved in the absence of humans. The two main species cultivated for coffee beans, *Coffea arabica* (the source of arabica beans) and *C. canephora* (the source of robusta beans), are both native to the highlands of Ethiopia but are now grown all over the world in similar climates.

As a PhD student, I used to buy coffee beans from a St. Louis roaster called Kaldi's Coffee. It was there that I learned about Kaldi, the apocryphal ancient goat herder mentioned in a 1671 treatise on coffee by the Maronite chronicler and professor Antoine Faustus Nairon. He wrote of a "certain camel herder or, as others say, of goats" from "Arabia Felix," who had complained to the monks that he had been awoken at night by his goats, which seemed to be "jumping."

Nairon explained that the prior of the monastery decided to find out why. When he investigated, he found the goats eating the berries of the coffee plant. A potion he made from the boiled beans gave him insomnia, and so he then ordered the rest of the monks to drink it to help them stay awake during the night watch and evening prayers.

We don't know for sure when coffee was first cultivated by humans. However, it was widely used in the Arabian Peninsula in antiquity, coming to Yemen through the Yemenite Sufi community around the fourteenth century BCE. But coffee didn't make it to Europe until the early 1600s. The Dutch began cultivating it in glasshouses in Amsterdam by 1616 and then in plantations in the East Indies. Thereafter came cultivation by the French, Spanish, and British in their own colonies.

The root of the word *coffee* is traced to the Arabic *qahwah,* which may have originally referred to a kind of wine but has the root *qahiya,* which means "to have no appetite." So the dark, red-wine-like color of coffee coupled with its appetite-suppressing effect are embodied in its name.

We didn't breed plants to produce caffeine, although we've helped create different modern cultivars from their wild relatives through artificial selection. These and all other caffeine-producing plants were making caffeine tens of millions of years before any humans were walking the earth.

On October 7, 1984, a breakthrough was announced in the *New York Times:* "Caffeine Is Natural Insecticide, Scientist Says." As the headline suggests, biologist James Nathanson had just demonstrated that caffeine is indeed a potent natural insecticide. Nathanson discovered this by incorporating powdered tea leaves and coffee beans into artificial caterpillar food, which he then gave to newly hatched tobacco hornworm caterpillars, which don't feed on plants containing caffeine in the wild. The hornworm adult is also called a hawk moth.

What Nathanson found shocked the world (although, by now, probably not you): "At concentrations from 0.3 to 10 percent (by weight) for coffee and from 0.1 to 3 percent for tea, there was a dose-dependent inhibition of feeding associated with hyperactivity, tremors, and stunted growth. At concentrations greater than 10 percent for coffee or 3 percent for tea larvae were killed within 24 hours."

Further experimentation revealed that the level of caffeine naturally found in undried tea leaves (0.68 to 2.1 percent) or undried coffee beans (0.8 to 1.8 percent) was sufficient to kill all of the caterpillars. Nathanson found the same insecticidal effects of caffeine on mosquitoes, beetles, butterflies, and true bugs, including at concentrations found in nature.

His most telling experiments involved spraying a mixture of caffeine on tomato leaves, the typical host plant for hornworms. Tomatoes don't produce caffeine, so these experiments were designed to mimic what the sudden evolution of caffeine might do to an herbivore that found itself eating a caffeine-producing plant. As the concentration of caffeine went up, there was a concomitant reduction in the amount of leaf chewed by the caterpillars. In other words, the caffeine protected the plant from attack by the hornworms.

A similar effect was found in 2002, when scientists in Hawaii accidentally discovered that a caffeine solution being tested as a toxicant to control the coqui, an invasive frog introduced from the Caribbean, also killed most of the large slugs found in their field plots. The researchers followed up by spraying or dipping vegetables in solutions containing caffeine concentrations of 1 to 2 percent, the same levels found in coffee beans, and offering them to the mollusks. Most of the snails and slugs died. And at far lower concentrations (0.01 percent), caffeine deterred them from feeding.

Although caffeine mimics the insecticidal effects of coffee or tea when sprayed artificially on plants that don't make it, this surface application is quite artificial. After all, the caffeine is made naturally inside coffee and tea plants' cells. Another way to sort it out would be to endow a plant species that does not normally synthesize caffeine in its cells with the ability to make it and see how resistant its leaves become.

Biologists did just that by genetically engineering caffeine production in tobacco plants, which don't normally make caffeine. The researchers spliced three caffeine-producing genes from the coffee plant genome into the tobacco plant's genome in the laboratory.

These transgenic tobacco plants produced levels of caffeine similar to those found in coffee plants. Leaves from tobacco plants carrying coffee-plant caffeine genes and control leaves without these genes were fed to tobacco cutworm caterpillars. The leaves producing caffeine were 99.98 percent less susceptible to herbivory than were the control leaves.

In naturally occurring caffeine-bearing plants like citrus, coffee, and tea, genes encoding the enzymes used to make caffeine evolved from existing genes that had performed a different function. Although we cannot board a time machine to determine why caffeine first evolved in any plant, the only known function of caffeine in plants is as direct or indirect defense against natural enemies. This role seems obvious, thanks to Nathanson's experiments. But caffeine might have first evolved as a molecule the plant used to signal the presence of stressors, just as salicylic acid serves as a hormone that signals the presence of attackers, rather than as a defensive strategy. Under this model, only later did caffeine become co-opted by plants as a toxin, just as willows took the ubiquitous plant-signaling hormone salicylic acid and turned it into a toxin by making much more of it.

Not surprisingly, some specialist herbivores of coffee plants have evolved the capacity to resist the toxic effects of caffeine. The most troublesome is the coffee berry borer, a small beetle that tunnels into the fruit and lays eggs in the beans. The larvae of this beetle consume the bean from within, rendering it unusable for coffee bean production. The beetle is native to the same African regions that gave rise to the two *Coffea* species now cultivated in the worldwide tropics.

The coffee berry borer has found coffee plants wherever they are, even in Hawaii, costing producers at least $500 million per year in damaged plants globally. Coffee is second only to petroleum in value as a global commodity, valued at $83 billion per year, and the borer is a major threat. The beetle itself cannot tolerate the insecticidal caffeine in its food. To get around it, the insect relies on enzymes produced by the bacteria living in its gut—its microbiome—which detoxify the caffeine.

When these beetles were treated with antibiotics that killed the bacteria and given their normal diet of coffee beans, they perished just as any other insect fed this diet would. The amount of caffeine consumed by one of these beetles in a meal is equivalent to the amount you'd find in five hundred cups of coffee, ten times the level that killed twenty-one-year-old Lachlan Foote in the early morning hours on New Year's Day in 2018.

Foote's tragic death in Australia was caused by an accidental caffeine overdose. He'd added one teaspoon of pure caffeine powder to a protein shake. After saying good night to his parents, he was found by his father in the bathroom, where he had died. It is not clear where he obtained the caffeine, but there was no warning label found on the bag he had used to store it.

Foote consumed at least five thousand milligrams of caffeine, equivalent to fifty cups of coffee. The FDA's recommended daily allowance for adults is four hundred milligrams. In the wake of his death, the family successfully pushed the Australian government to ban food additives with caffeine concentrations above 5 percent and liquids with levels above 1 percent. A ban went into effect less than a year later.

In October 2014, I was out of town when I learned that my dog was in critical condition at an emergency veterinary hospital in Tucson. She had broken into the dog sitter's backpack and torn into a bottle of gelatin-coated caffeine pills.

The dog ate more than the eight to thirteen pills needed to achieve caffeine toxicosis, given her size. The dog sitter desperately tried to save her and, with the help of my neighbor Erika, brought her to the vet. My dog fell into a coma despite their heroic efforts.

I raced home. The vet suggested that we prepare ourselves that the dog might not survive. But under their steady care, and with a lot of luck, she began to awaken, weak and confused but with her personality intact.

These dark lessons teach us that plants don't make caffeine for our benefit. They make it as a defense against being eaten. However, we've learned to use caffeine to improve our own lives because, unlike the tiny bodies of insects, our own big bodies can handle fairly large doses of this alkaloid and because our brains seem prewired to interact with it.

Wired

Most of us don't drink coffee or tea for its long-term potential to extend our lives or to prevent age-related neurodegeneration. We drink it to wake up, hone our focus, enhance our memory, quicken our step, lift our mood, and push ourselves through the winter doldrums and the more difficult times of our lives. Being wired makes most of us happier. Athletes and mathletes alike use caffeine for its performance-enhancing effects on the body and the brain.

Caffeine's influence on our mood may be more than superficial. Caffeinated coffee was associated with reducing the risk of depression by 25 to 50 percent in studies involving hundreds of thousands of people, and the risk was further reduced by each additional cup consumed, up to a point.

Drinking two or three cups of caffeinated coffee per day is associated with a staggering 45 percent reduced risk of suicide compared with those who don't drink it. Bump that up to four cups per day, and the odds of suicide decrease by 53 percent. However, eight or more cups a day increase suicide risk by 58 percent. Again, we see U-shaped protective effects for both depression and suicide risk, a common pattern with so many of nature's toxins when we use them as drugs.

Although coffee has a reputation for exacerbating heart rhythm abnormalities, or arrhythmias, this indictment may be unwarranted. The largest study to date, of more than 350,000 people, showed that each cup of coffee consumed *lowered* the risk of any type of heart arrhythmia, including atrial fibrillation, supraventricular and ventricular tachycardias, and premature atrial and ventricular complexes.

However, a more recent study used text messages to instruct one hundred patients to either consume or avoid caffeinated coffee across fourteen days. Patients wore a device that measured the heart rhythm. Premature ventricular contractions increased by 50 percent in those who drank more than a cup a day. For most of us, this temporary side effect is not a health concern.

But not everyone should be drinking a lot of caffeinated coffee. Caffeine poses clear downsides for people with anxiety, eating disorders, and sleep disorders. It can interfere with some medications, and its consumption needs to be carefully managed in pregnancy, as discussed earlier.

There are, of course, people who simply don't like the way caffeine makes them feel. Studies involving hundreds of thousands of people in the

United States and the United Kingdom have attempted to ask if our preferences, consumption rates, physiological sensitivities, and ability to metabolize caffeine could be explained at least partly by slight differences in our genetic codes. The answer to all these questions is yes. An important fraction of the variation between some of us in these traits may be controlled by different versions of genes involved in smell and taste, the brain's response to caffeine, and the body's ability to detoxify it.

Roughly one-half of the overall differences in caffeine consumption patterns is heritable, at least in the people living in the geographic regions where this trait has been measured. If we can extrapolate the whole of humanity, which is a big if, this heritability means that we largely resemble our biological relatives in caffeine consumption habits.

In studies involving female twins, the correlation between their caffeine use is even higher, rising to 77 percent. How well we tolerate caffeine, and the severity of our withdrawal symptoms, are also heritable, although to a lower degree—between 35 and 45 percent of the variation among people in some populations.

While some of the variation in traits between individuals depends on the particular set of genetic variants inherited from their biological parents, an even larger fraction of the variation in coffee consumption patterns between people is determined by culture and environment.

Caffeine is perceived as bitter by animals that taste it, including us. This experience in our mouths is mediated by taste receptors encoded by a set of *TAS2R* genes. Some of us are more sensitive to bitter substances than others are, and this sensitivity partly depends on which variants we carry at these *TAS2R* genes. A substantial fraction (36 to 73 percent) of the variation between us in our bitter sensitivity is determined by the specific set of *TAS2R* variants we inherited from our parents.

In a large study within the United Kingdom, geneticists found a telling relationship between an individual's coffee consumption and how intensely they perceive three substances: caffeine, a bitter chemical called propylthiouracil, and quinine. Those who were more sensitive to the bitterness of propylthiouracil and quinine consumed less coffee, and vice versa.

Why the difference? On the one hand, most toxins are bitter, so an innate aversion to bitterness is often advantageous to people and other animals. Those with higher propylthiouracil and quinine sensitivities are more averse to foods containing them and, accordingly, to coffee. Surprisingly, on the other hand, those with heightened caffeine bitter sensitivity do the opposite and consume more coffee.

Some of you probably like bitter drinks and foods. But I would bet that the bitter ingredient is paired with a rewarding one, like a sweet taste, in the case of tonic water, which has sugar added along with the quinine.

To figure out what is really going on, the researchers found that the *TAS2R* genetic variant associated with higher sensitivity to caffeine was more strongly associated with increased intake of caffeinated than decaffeinated coffee. This association explained, biologically, the findings about higher bitter sensitivity being related to higher coffee intake from the survey. The conclusion — that those of us with heightened sensitivity to the bitter taste of caffeine learn to associate it with psychostimulatory reward — is a remarkable and counterintuitive finding.

One of the ways that caffeine works its magic in our brains is by binding to the adenosine A_1 and A_{2A} receptors in our nerve cells. The neurotransmitter adenosine's most salient role is as a modulator of sleep. Its levels increase in the brain as a function of how long we are awake; a certain level of adenosine eventually induces sleep.

Caffeine is a molecular mimic of adenosine, but it has the opposite effect on sleep because it binds to adenosine A_1 and A_{2A} receptors, thereby blocking adenosine. This binding activity is responsible for caffeine's ability to arouse us. Adenosine can't do its job when caffeine blocks the adenosine receptors in our brains.

Some of us carry rarer but natural mutations in our adenosine A_{2A} receptor genes. These changes lead to increased or decreased sensitivity to caffeine, which corresponds with its anxiety-inducing effects and habitual use. One of these adenosine A_{2A} receptor gene variants is associated with avoidance of high-caffeine drinks and dark chocolate. Those of us with this variant may tend to avoid coffee more than do those without it, because we perceive coffee as bitter and are more likely to drink tea, which has half as much caffeine as coffee.

Other people have a different variant at the adenosine A_{2A} receptor gene. This natural mutation is associated with an increased fondness of unsweetened coffee, increased intake of coffee, low aversion to the bitter taste of coffee, and greater enjoyment and consumption of dark chocolate.

The speed with which we detoxify caffeine, once it has entered our bodies, and remove it from our blood is determined primarily by the CYP1A2 enzyme in our livers. Mutations affecting this enzyme are among the most important when it comes to explaining variations in the speed at which individuals metabolize caffeine.

Like the adenosine A_{2A} adenosine receptor variants, these CYP1A2 variants are also associated with whether we drink caffeinated beverages, how much of them we drink, and in what form. One CYP1A2 variant is associated with higher coffee intake, higher consumption rates of caffeinated than of decaffeinated coffee, higher total tea intake, greater liking of unsweetened coffee, and greater liking and intake of dark chocolate than the other variants. The bottom line is that those of us with less sensitive adenosine receptors and more efficient caffeine metabolism consume more coffee or tea to reap the same positive effects of caffeine than do people with more sensitive and less efficient caffeine metabolisms.

Even more intriguing is the possibility that we may condition our response to the bitter taste of caffeine according to how our brains interact with it and how quickly our livers detoxify it. This means that we can change our innate aversion to bitter dietary chemicals like caffeine if the reward is high enough. In other words, we alter our behavior—and tastes!—through caffeine-induced reward. The degree to which we learn to like caffeine-containing food and drink relies in part on the genetic variants that independently influence our taste perception, our sensitivity to caffeine's effects on our brains, and our body's ability to detoxify caffeine efficiently.

Besides caffeine, dark chocolate, but not white, contains chemicals identical to our brain's own endocannabinoids, including anandamide, an endogenous cannabinoid chemical found in marijuana. Both anandamide and delta-1-tetrahydrocannabinol (THC) bind to our body's cannabinoid receptors and produce a euphoric feeling. In addition, higher anandamide levels in the brain produce antidepressant effects, at least in rats. Although the levels of anandamide in chocolate may not be high enough to actually influence the brain on their own, two related molecules in chocolate slow the breakdown of the body's endogenous anandamide levels. So, consuming chocolate could still elevate our cannabinoid levels, although the extent of this potential effect has not been studied. The dearth of research notwithstanding, the mental and physical lift I get from chocolate's caffeine and related alkaloids, not to mention the sublime taste of chocolate, is yet another reason to enjoy it.

Beyond many humans' liking of caffeine—and most insects' avoidance of this substance—caffeine plays another attracting role. Surprisingly, some plants use this alkaloid to appeal to insects.

Dancing Bees and Caffeinated Flowers

Citrus species and *Coffea* species, whose native ranges overlap with those of honeybees in Asia and Africa, naturally produce caffeine in the nectar of their flowers. In 2013, researchers discovered that this was no coincidence. When honeybees gathered nectar at caffeinated flowers, the bees have enhanced long-term memories of the flower's scent, thanks to the caffeine in the nectar. So, in practical terms, the bees would not only be more attracted to other flowers producing that scent but would also more enthusiastically recruit their sister worker bees to gather nectar from the same flowers.

The levels of caffeine in the nectar are below the threshold of bitter detection by honeybees and below the toxic levels that Nathanson measured in tea leaves and coffee beans. The implication is that the flowers produce caffeine in the nectar as a way to manipulate the insect brain, producing levels just low enough to get past the bitter taste receptors that all animals have but still high enough to manipulate the bee brain. The plants are not consciously plotting, of course, but natural selection has endowed them with the ability to make just the right amounts of caffeine to be effective at enhancing their fitness.

Before we get to the bottom of this sweet subterfuge, we must first understand how honeybees gather nectar for honey production. Honeybees have evolved to be among the most sophisticated social animals because they are able to do something you or I could never do. A single honeybee forager can communicate to her sisters the exact location of a new flower patch she has found by doing a unique waggle dance in the darkness of the hive. The other bees then use this information to fly to the patch without ever having been to it before.

Neuroethologist Karl von Frisch won the Nobel Prize in Physiology or Medicine in 1973 for discovering how honeybees do this, which is a wonder of the natural world. The dance language of the honeybee isn't the whole story. In practice, the scent of the flowers that the forager has carried back to the hive is also needed for her sisters to find the right patch of flowers. That's where the caffeine-laden nectar comes in.

As noted earlier, when we consume caffeine, the molecules block adenosine receptors in our brains. But what I didn't tell you was that the blocked receptors cause the neurons to release neurotransmitters that help create long-term memory. The enhanced odor-associated memory effect provided by caffeine in the nectar works the same way in bees. It seems quite likely that the plants are dosing their nectar with caffeine to help manipulate bees into returning to pollinate them.

When researchers placed sugar at bait stations laced with caffeine, honeybee foragers wildly overestimated the quality of the food resource, compared with bees that fed on caffeine-free sugar. A stimulant-induced rave-like atmosphere erupts in the hive, enticing four times as many bees back to the caffeine-laced bait. The artificially caffeine-spiked bait induces inefficient foraging and, potentially, reduced honey stores.

By adding caffeine to nectar, a plant's reproductive success may be higher than it would be with just sugar and amino acids as a reward, because the bees will move more pollen around. But this lure of caffeine may not be all good news for the bees, given all the extra energy expended for naught.

Rather than creating a win-win, the plant, by drugging its pollinators, may obtain the benefit at the expense of the honeybee colony. This degree of manipulation goes a step further than the odor-driven deceptive pollination systems used by corpse flowers and Solomon's lilies or the way that bitter aloe nectar attracts birds instead of bees.

Like caffeine, high levels of nicotine evolved in some plants as an insecticide. Then, Indigenous peoples of the Americas began using nicotine some twelve thousand years ago. After Big Tobacco corrupted its use, tobacco became, by a large margin, the biggest killer among all of the natural products that we use. I'll expand on the story of nicotine in the next section.

Nicotine and Its Natural History

Nicotine is an even more potent natural insecticide than caffeine, protecting tobacco plants from herbivore attack. The first study on its use as an insecticide was published in 1916 by the US Department of Agriculture (USDA).

So potent is nicotine as a neurotoxin that a disease called green tobacco sickness once ran rampant in tobacco harvester workers. Symptoms include headache, nausea, vomiting, dizziness, and prostration on the ground. In a study in North Carolina, researchers found that nicotine was present at high concentration in the morning dew on the tobacco leaves and was probably directly absorbed into the skin of the workers.

More recently, scientists have developed so-called neonicotinoids, synthetic insecticides that use the nicotine molecule as foundation. These neonicotinoids are problematic because they can harm not just crop pests but also nontarget insects, like bees, that interact with the plants. What's more, bees can become "addicted" to nectar containing these neonicotinoids, much like the bees that are more faithful to flowers with nicotine in their nectar. Ultimately, researchers have shown that for bumblebees, neonicotinoid exposure makes them poorer caretakers and nurses of the bees that they are rearing back at the nest.

Yet, as you may have guessed, there are some herbivorous insects that have evolved to become specialized on tobacco plants. Among the most familiar to many people is probably the tobacco hornworm, which attacks tomato, tobacco, and sacred datura plants—all members of the Solanaceae family.

Not only has the tobacco hornworm evolved mechanisms for coping with nicotine, but it also uses the toxin as a defense against predatory spiders. In Utah, Pavan Kumar and colleagues found that tobacco hornworms that attack coyote tobacco use an enzyme called CYP6B46 to transport the nicotine from their gut to the blood. From the blood, nicotine is released into the air through the caterpillar's spiracles, the breathing holes that dot the sides of its body. The volatile nicotine concentrations released by the spiracles are high enough to reduce spider attack rates.

The researchers who discovered the smoker's breath of the tobacco hornworm coined the term *defensive halitosis*. Anybody who has kissed a smoker knows how unpleasant the experience can be, all things considered, and spiders apparently feel the same about their own close encounters of the tobacco hornworm kind!

If this weren't enough, the nicotine in the blood of the hornworm ends up harming the larvae of parasitoid wasps that hatch inside the bodies of the hornworms. In the breath and in the blood of tobacco hornworms, nicotine serves as a chemical defense from free-living enemies and the enemies within.

Human use of tobacco began as early as 12,300 years ago, according to recent evidence unearthed from Utah, a few hours' drive north of where the tobacco hornworms that repelled the spiders were studied. Previously, the oldest evidence of tobacco use by Indigenous peoples of the Americas came from pipe residues dating to around 3,300 years ago. As scientists did with "Sid" the Neanderthal, researchers can recover nicotine from the tartar of human teeth discovered in the archaeological record. Tartar from both an Indigenous woman buried around 630 years ago and a man buried 420 years ago not far from where I live in what is now Oakland was positive for nicotine.

The hair of a baby who died 2,400 years ago in the highlands of the

nicotine

Atacama Desert of Chile was also found to contain nicotine. Remarkably, the researchers determined that the nicotine came through the placenta rather than from the mother's milk because they were able to look at its deposition as the hair grew in utero. This passive consumer may have been born addicted to nicotine, although the researchers speculate that given the high levels of nicotine found in the mother, she may have miscarried. Such high levels also suggest that the mother was a tobacco shaman.

Independently, tobacco in the form of pituri—a mixture of tobacco leaves and a specially prepared ash—has been used for millennia by Aboriginal Australians living primarily in central Australia. Pituri is still used today in Australia, especially as a quid, although it is also smoked. When chewed, the leaves are mixed with ash and rolled by the tongue into the buccal cavity between the cheek and gum to release the nicotine and related nornicotine.

A variety of tobacco species, including members of the *Nicotiana* genus and, more importantly, the related plant *Duboisia hopwoodii,* from the Solanaceae family, are used in pituri preparations.

Pituri is traditionally used as a stimulant, as a trance-inducing drug as it was in the Americas, and to stave off hunger on long treks. There is even a report of pituri being used to poison prey, as noted in this 1899 account, although the veracity of the description has not been substantiated: "Leaves of the pituri plant (*Duboisia Hopwoodii*) are used to stupefy the emu. The plan...is to make a decoction in some small waterhole at which the animal is accustomed to drink. After drinking the water the bird becomes stupefied, and easily falls a prey to the...spear." Pituri was valuable in trade, was bartered across Aboriginal trade routes, and was a potent status symbol.

Remarkably, humans employed nicotine completely independently in both Australia and the Americas for thousands of years for a variety of practical and spiritual purposes. This convergent cultural evolution occurs for many other psychoactive chemicals, including ethanol, cardiac glycosides, and, as we will soon see, alkaloids found in water lilies. Like the many distantly related animals (monarch butterflies and milkweed beetles, for example) that concurrently evolved to attack the same toxic plants, distantly related human societies do the same.

Nicotine and Nitrosamines

Pure psychoactive drugs, most of them alkaloids, have become available in the blink of an eye, from the perspective of human evolution. If these pure forms of the drugs were not so easy to obtain, deaths from overdoses, which usually stem from a drug use disorder, would be virtually eliminated.

In the United States, nearly 92,000 people died in 2020 because of a psychoactive drug overdose, most of the deaths unintentional. Almost 70,000 of these deaths were related to use of natural, semisynthetic, or synthetic opioids, with the remainder caused mostly by amphetamines and cocaine. Although alcohol is typically not imbibed in pure form, abuse of refined ethanol in alcoholic drinks kills an additional 95,000 people per year in the United States. Around half of these deaths, like my father's, are attributed to indirect effects of cancer, cardiovascular disease, and liver disease. In total then, what we might call the use of refined drugs kills around 200,000 people per year in the United States, mostly through overdoses but also because of chronic disease.

However, these numbers pale in comparison to deaths due to tobacco use. In the United States, 34.1 million people smoked tobacco products in the year 2019, and around 500,000 US deaths were attributed to tobacco that year. Zooming out, 1.14 billion people smoked tobacco products globally in 2019, for a total 7.7 million tobacco-related deaths that year. Almost all who died (87 percent) were current smokers, and of course the deaths are nearly all unintentional. These statistics lie on top of the 200 million disability-adjusted life years lost per year.

Of those who died of any cause in 2019, roughly 20 percent of men and 6 percent of women died from smoking-related causes. This comports with the estimate that nearly two-thirds of long-term smokers will die of smoking-related diseases. Although there has been a 28 percent drop in tobacco use by men and a 38 percent drop by women globally since 1990, population growth nevertheless increased the total *number* of people who use tobacco. Tobacco use remains the single biggest mortality risk factor for men.

One of my favorite photos of my dad was taken when he was in the navy during the Vietnam War. The draft board told him that he was going to be drafted, so rather than being drafted into the army, he enlisted in the navy. He wanted to be a physician and decided that being a corpsman was most aligned with that goal. Little did he know that the corpsmen of the navy were the medics for the marines, who were on the front lines in Vietnam.

He trained in Southern California with the marines, and during one of the exercises, a friend took a photo. My dad's face was painted green, his helmet decorated with vegetation for camouflage, a cigarette perched in his mouth, and a rifle in his hands. He smoked in his twenties but was somehow able to quit by the time my brother and I were born.

Just before I started my job at Berkeley, a series of stressful events got the better of me, and I began to secretly smoke cigarettes — again (I had smoked in college). I smoked just one a day, which, I told myself, was fewer than the number of cigarettes that then president Barack Obama said he smoked some days in the White House.

The smoke from the first few cigarettes stung my throat, lungs, and eyes. Once the coughing subsided, the euphoria that ensued from the firing of that dormant dopamine-releasing brain circuit was superlative. A quiet walk each night around the neighborhood, just me and my one cigarette, provided exactly the mental shift I wanted as these big, stressful events in my life were unfolding.

Of course, I knew perfectly well that this old friend I'd reconnected with could kill me. Even just one cigarette a day raises the odds of stroke or heart attack, and in the longer term, aortic aneurysm, chronic obstructive pulmonary disease (COPD), and cancer risk. I also understood that nicotine, rather than lessening my anxiety, actually increases it, even if it does have other positive cognitive effects. Still, I continued to smoke for a few months, and I thoroughly enjoyed it, even if I knew deep down that smoking was a mistake. And I knew I would feel ashamed if anybody found out. As I puffed away, I wondered whether it was the nicotine itself or other chemicals in tobacco that cause these deaths.

A major difference between tobacco and the pure forms of alkaloids like amphetamines, cocaine, opioids, and, to a lesser degree, diluted ethanol, is that mortality from nicotine overdose itself is essentially negligible. The mortality caused by tobacco use, primarily smoking it, arises from cancer, respiratory diseases like emphysema, and cardiovascular disease resulting from the more than nine thousand chemicals and sixty-nine known carcinogens produced by the combustion, or smoke, of tobacco leaves. Nicotine itself is thought to play a relatively small role in illnesses related to tobacco use, although our knowledge in this area is changing rapidly.

Does this mean nicotine is safe? The right answer is important. Around 15 percent of teenagers in the United States and over 3 percent of adults now consume nicotine by vaping with e-cigarettes. Moreover, nicotine is also chronically consumed through gums, lozenges, and patches.

The evidence that does exist, while scant, suggests that in otherwise healthy adults, especially those who haven't damaged their cardiovascular systems already, nicotine poses a lower risk of causing cardiovascular problems than does tobacco smoke. However, some evidence suggests that heart cells, at least in laboratory experiments, can be damaged by the breakdown products of nicotine.

Vaping e-cigarettes significantly increases the risk of asthma, chronic bronchitis, emphysema, and COPD in the United States. The nicotine might explain the increased risk of respiratory disease in e-cigarette users. Mice exposed to nicotine-containing e-cigarette vapor for one hour per day for four months developed COPD-like symptoms, whereas mice exposed to vapor without nicotine did not.

In addition to nicotine, e-cigarette aerosol contains toxicants like aldehydes, acrolein, and metals. These toxic substances may cause acute injury to the types of cells that line the blood vessels and cause oxidative stress. The other toxins present probably depend on the particular solvents and flavorings added, the temperature of ignition, and the composition of the vaping pen used.

The longer-term effects of these and other chemicals in e-cigarettes on cardiovascular risk and cancer risk are simply unknown. The outbreak of acute lung injury that peaked in 2019–2020 in the United States was not associated with nicotine from e-cigarettes. The culprit instead was the THC-containing e-cigarette products and the presence of vitamin E acetate in the lungs of those affected.

We know little about the impacts of vaping on public health, and questions about the impacts get more complicated if we look at nicotine as a carcinogen. There is evidence that long-term nicotine use in mice, rats, and human cell lines can lead to cancer.

One of the more convincing studies exposed mice to nicotine vapor from e-cigarettes. It found that after a year, nearly one-quarter of the mice exposed to the nicotine vapor had developed adenocarcinoma tumors in the lungs. None of the mice exposed to control vapor without the nicotine developed these malignancies.

A similar pattern was seen for bladder hyperplasia, an overgrowth of the bladder lining that can lead to cancer. Over half of the mice exposed to nicotine vapor developed hyperplasias, compared with only one exposed to the control vapor and none exposed to filtered air.

The mechanism driving the association between nicotine and cancers is likely to center on the mammalian body's inability to convert all the nicotine into cotinine, which is then excreted in the urine. Some of the nicotine breaks down into nitrosamines, including nitrosamine ketone, which can further break down into methyldiazohydroxide, a chemical that directly binds to DNA in our cells and causes harmful mutations.

Mutations in the DNA molecule are the direct causes of most cancers. Other nicotine breakdown products may also prevent the cell's DNA repair machinery from repairing these mutations. Although nitrosamines are 95 percent lower in e-cigarette smoke than in tobacco smoke, there is concern that this knowledge has led the public to believe that the cancer risk is minimal.

Because other studies, however, have shown no increased cancer risk posed by nicotine itself, the carcinogenic role of nicotine has been controversial. As you can surmise, the jury is still out, and the answer may well be quite nuanced and may depend on many environmental, genetic, and other individual factors.

Despite the negatives, evidence from more than seventy observational studies suggests that the use of nicotine is protective against Parkinson's disease. This disease is caused by the death of particular brain neurons that produce dopamine. When nicotine enters our bodies, it binds to the nicotinic acetylcholine receptors; this action gives rise to nicotine's euphoric, dopamine-releasing effects, including the stimulatory effect brought on by the activation of the mesolimbic reward system. Nicotine's ability to stimulate the neurons expressing the nicotinic acetylcholine receptors may be one direct route through which this alkaloid protects against the disease. Another possible route is that when nicotine binds to these receptors, it may trigger a chemical cascade that ends up shielding against inflammatory-induced and oxidative-stress-induced injury of the brain's nerve cells.

A small clinical study that used nicotine patches to determine whether the progression of Parkinson's disease could be slowed found no effect. So, nicotine perhaps has more of a preventative role than a therapeutic one. However, another small clinical trial found that wearing a nicotine patch improved attention and memory in adults with mild cognitive impairment.

Remarkably, increased consumption of foods that contain nicotine, like plants from the Solanaceae family, including tomatoes, potatoes, and especially peppers, is also associated with reduced Parkinson's disease risk, although the nicotine itself can't be isolated as the causal factor. Eating more solanaceous vegetables may confer a protective effect against it, but one would have to eat *a lot* in one sitting to match the amount of nicotine in one cigarette.

It isn't just the human brain that seems to perform better under the influence of nicotine. Just as experiments with caffeine and bees showed, bees that can overcome an initial aversion to this alkaloid and feed on nectar containing nicotine learn colors faster and return to these flowers more faithfully than to flowers that produce nectar without nicotine, even if the nectar reward is artificially reduced over time.

As described, when humans and other mammals consume nicotine, an enzyme in the liver detoxifies it into cotinine and then 3-hydroxycotinine. Two genes, one for the nicotine receptor and one for the nicotine detoxification enzyme CYP26A, exist in slightly different forms in different people. Mutations in the DNA within these genes occurred long ago in some of our ancestors and were passed down generation after generation.

Significantly, some of these genetic variants clearly predispose some people to cancers and other illnesses arising from tobacco use and possibly even from nicotine in e-cigarettes. A study involving thousands of Americans of European descent discovered that two mutations in and around the gene that encodes the nicotinic acetylcholine receptor subunit 5 could help explain why a given person is a heavy or light smoker. These two mutations can also help predict the risk of lung cancer. People carrying the rarer form of the receptor were more likely to smoke heavily and were also at higher risk of lung cancer, even when controlling for the amount smoked.

This study was repeated in Denmark. There, researchers found similar patterns: increased risk of lung cancer, bladder cancer, and COPD in people carrying one of the rarer variants. One explanation for this finding is that a larger fraction of people with the rare receptor variant also inhaled more smoke and smoked more per day over their lives than did those carrying the more common receptor variant.

The mutation in the rarer receptor variant renders the receptor less responsive to nicotine. So, because more nicotine is needed to gain the desired effect in the brain, people with this gene mutation may tend to smoke more.

The speed with which our bodies remove nicotine from the bloodstream is also a key piece of the puzzle. A study of thousands of smokers in Finland found that the ability to metabolize nicotine into cotinine was highly heritable. In other words, if the parents were fast metabolizers, the children tended to be, and vice versa for slow nicotine metabolizers.

The genomes of the study's participants were scanned to determine which, if any, genetic differences between people might be correlated with different rates of nicotine metabolism. The single biggest difference was in a gene encoding the liver enzyme CYP2A6, which detoxifies nicotine. One version of this enzyme causes faster metabolism of nicotine, and other versions cause slower metabolism, much as different forms of the CYP1A2 determine our ability to detoxify caffeine.

People with the fast version of the CYP2A6 enzyme smoked more and liked smoking more than did slower metabolizers. A remarkably similar pattern is observed with people who have the fast version of CYP1A2 and have increased coffee intake.

Fast nicotine metabolizers were also less likely to be successfully treated with nicotine replacement therapy relying on nicotine patches than slow metabolizers were. The fast metabolizers were too efficient at removing nicotine from their bodies for the patch to provide the fix they sought.

In contrast, a replacement therapy using a drug called varenicline instead of nicotine was effective in these fast nicotine metabolizers. Because varenicline is not detoxified by the CYP2A6 enzyme, the drug was more effective than nicotine-replacement therapy for the fast metabolizers. These findings hold great promise for the health of smokers. Because fast nicotine metabolizers smoke more tobacco, they are at greater risk for cancer and tobacco-related diseases. It is harder for fast metabolizers to quit if replacement therapy uses nicotine; using varenicline may be more effective.

The parallels between both sensitivity to nicotine and caffeine at the receptor level and people's nicotine and caffeine consumption patterns, including how much the users like these stimulants and how much they inhale, or drink, are uncanny and point to a more general observation: we are prewired to interact with these stimulants, and our genetic predispositions help determine the nature of our relationship with them.

When it comes to caffeine and nicotine, like most of nature's toxins we use, the line between poison and cure is thin. That's also true for the alkaloids scopolamine, cocaine, and curare. No three drugs have played a more important role in the development of modern surgery than they have. And again, as I discuss in the next chapter, we owe our use of these remarkable substances to Indigenous knowledge holders.

8.

Devil's Breath and Silent Death

Cleopatra: Charmian!
Charmian: Madam!
Cleopatra: Ha, ha!
Give me to drink mandragora.
Charmian: Why, madam?
Cleopatra: That I might sleep out this great gap of time
My Antony is away.
Charmian: You think of him too much...
Cleopatra: O Charmian,
Where think'st thou he is now? Stands he, or sits he?
Or does he walk? Or is he on his horse?
O happy horse, to bear the weight of Antony!
Do bravely, horse! For wot'st thou whom thou movest?
The demi-Atlas of this earth, the arm
And burgonet of men. He's speaking now,
Or murmuring "Where's my serpent of old Nile?"
For so he calls me: now I feed myself
With most delicious poison.
> —WILLIAM SHAKESPEARE, *ANTONY AND CLEOPATRA*

Scopolamine and Twilight Sleep

We've all been there. Seemingly out of nowhere, cupid's golden-tipped arrow hits us. When it strikes, we fall in love and surrender to an uncontrollable, preprogrammed release of hormones and neurotransmitters. The feeling can be overwhelming—even unwanted. And if that love is unrequited or if the object of desire is suddenly snatched away, it can be unbearable.

Shakespeare's Cleopatra is so lovesick for Mark Antony that she instructs her lady-in-waiting Charmian to bring her the mandrake tonic, her "most delicious poison" to stem the tide. That phrase embodies the double-edged sword that is part of the bargain when it comes to our use of nature's toxins.

In medieval Europe, there may have been no better way to check out than to drink of the mandrake—an essential ingredient in any witch's brew. The use of the mandrake as a soporific or an amnesia-including tonic was due to the presence of tropane alkaloids like scopolamine. Scopolamine produces "drowsiness, euphoria, amnesia, fatigue, and dreamless sleep."

This alkaloid is one of the most potent psychedelic compounds in nature's pharmacopoeia but has many other powers too and is still used as a medicine. Also called hyoscine and colloquially as the street drug devil's breath, scopolamine is on the World Health Organization's *Model List of Essential Medicines*.

Scopolamine is an anticholinergic drug; it blocks the muscarinic acetylcholine receptors. When these receptors are blocked, nausea is prevented. Accordingly, one of scopolamine's more common medical uses now is in the form of a slow-release patch behind the ear to prevent motion sickness (sold as Transderm-Scop). Modified forms of scopolamine include scopolamine butylbromide (sold as Buscopan) and methscopolamine (Extendryl, AlleRx, Rescon, or Pamine).

People around the world use the mandrake and other plants in the Solanaceae as an entheogen—a drug for spiritual practice—for the plants' scopolamine content. Plants that make scopolamine include some of those in the genus *Datura* and are variously named devil's trumpet, devil's snare, jimsonweed, moonflower, sacred datura, and thorn apple. Georgia O'Keeffe's paintings of a datura flower captivated Shane and me when we visited the eponymous museum in Santa Fe.

Another important source of scopolamine from the Solanaceae are Neotropical plants in the genus *Brugsmansia*, collectively called angel's trumpets, or *borracheros* in Spanish. Some species of *Duboisia* in Australia make this important alkaloid as well. As we learned earlier, pituri is made by Aboriginal Australians from *Duboisia hopwoodii*, which also produces nicotine.

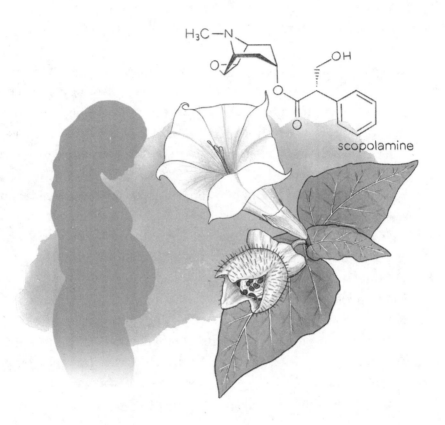

The use of scopolamine-producing plants as medicines and entheogens is so widespread across Indigenous cultures around the world that it seems impossible to know if its use arose many times independently or only once and then spread. Scopolamine is particularly important in the Americas, in parts of East Africa and across South Asia and East Asia.

I literally run into angel's trumpets every day when I walk to campus in Berkeley. Most species of these *Brugsmansia* plants are native to cloud forests in the northern Andes, which contain microclimates that are similar to those in the Bay Area, so they do particularly well here.

The huge white or yellow bell-shaped flowers dangling next to the sidewalks here are striking. Some are pollinated during the day by hummingbirds with incredibly long bills, and others by bats and moths at night. Yet, the seeds that develop in the ovaries of the fruits are protected from seed predators because of the high concentrations of scopolamine and other tropane alkaloids that lurk within.

The Zuni Pueblo in western New Mexico, despite the odds, still stands in the same place it stood when the Spanish first invaded. The Zuni people have lived there for three thousand to four thousand years, and their use of the sacred datura has been woven into many elements of life.

Historically, Zuni rain priests and members of select fraternities gave powdered datura root to render patients unconscious so that medical procedures could be performed. These operations included "setting fractured limbs, treating dislocations, making incisions for removing pus, eradicating diseases of the uterus and the like."

Such use presaged the entry of *Datura stramonium* into the Europe medical establishment in 1762 by Viennese doctor Anton von Störck, who included it in a pamphlet that described its potential medicinal uses. He also illustrated black henbane (*Hyoscyamus niger*), a relative with the same tropane alkaloids and widely used as both medicine and poison.

Scopolamine reduces salivation and other bodily secretions, prevents nausea and vomiting, and suppresses muscle spasms. These features are in addition to its soporific effect. The alkaloid therefore became a natural candidate for use in anesthesia before surgery, as the Zuni had used it for millennia. The use of *Hyoscyamus* was even mentioned in the Papyrus Ebers to treat "magic in the belly," which can easily be interpreted as a description of seasickness or an upset stomach.

Scopolamine got its name thanks to the Tyrolian-born physician Giovanni Antonio Scopoli, who collected and preserved a plant that was later described as *Hyoscyamus scopolia* and as *Scopolia carniolica* in his honor. The drug's use in modern medicine began in earnest in the early 1900s after a German physician proposed that it be used during surgery.

By 1902, Austrian physician Richard von Steinbüchel recommended that a mixture of scopolamine and morphine be given to mothers during childbirth because not only would the women feel no pain through the morphine but they would also not recall giving birth at all, thanks to the scopolamine. Eventually, two German physicians, Carl Glauss and Bernhardt Kronig, began to further explore the use of combined scopolamine and morphine during labor. *Dämmerschlaf,* or "twilight sleep," was introduced to the European medical establishment in 1906 during an obstetrics conference in Berlin.

Given that there was no generally safe way to reduce the pain of childbirth at the time, word began to spread. By 1914, the drug combination gained a foothold in the United States when an article titled "Painless Childbirth" appeared in *McClure's Magazine.* The article framed the use of scopolamine and morphine as a way for women to end their suffering during labor. Twilight sleep was swept up in the first-wave women's rights movement in the United States because it was seen as liberating women from the pain of childbirth. Many people in the United States considered the pain of childbirth a punishment exacted by God ever since Eve and original sin.

Initially, twilight sleep was seen as a progressive win-win. Doctors could still get their patients to respond to them during childbirth, but the patients wouldn't remember the conversation, nor would they feel the pain. However, it turned out to be difficult to get the doses of these two powerful alkaloids right, and a high-profile death of Charlotte Carmody during childbirth led to its demise. Ironically, Carmody was one of the proponents of twilight sleep for childbirth, and her death may have been unrelated to the drugs. Despite the challenges of dosing, twilight sleep was used sporadically in the United States through the 1960s.

Scopolamine even moonlighted as a "truth serum" in interrogations both in the United States and more widely behind the Iron Curtain in the mid-twentieth century. And in other places around the world, countless people, including those in many Indigenous cultures, have tapped the powers of scopolamine-synthesizing plants for myriad purposes.

Scopolamine derived from *Brugmansia* has also been used criminally, for example, during robberies and sexual assault, allegedly in major urban areas in Ecuador and Colombia. According to the US Overseas Security Advisory Council, victims are purportedly given a dose of scopolamine in food, drink, or even through the skin via residues on pamphlets. It is in these situations that scopolamine is called devil's breath. However, there is little evidence on exactly what lies at the root of the more than fifty thousand incidents claimed by the council in the late 2010s.

The rare toxicology analyses conducted on survivors of sexual assault have found well-known tranquilizers widely implicated as date rape drugs, rather than scopolamine. Scopolamine is manufactured by drug companies and is not easily extracted and purified from *Brugsmansia*. So, the use of scopolamine derived from the plants in these cases is questionable, despite the urban legend of the devil's breath, and the devil is in the details.

A different urban legend might hold more truth. Atropine is another important tropane alkaloid found in plants like *Brugsmansia* and the mandrake. This alkaloid is synthesized in high concentrations in the belladonna plant (*Atropa belladonna*), whose species name means "beautiful woman." In fifteenth- and sixteenth-century Italy, some women used belladonna as a cosmetic to dilate the pupils for a doe-eyed look. Atropine is now used to treat some kinds of myopia in children.

In the next section, we'll examine two other alkaloids important in medicine.

Coca and Curare

Two alkaloids essential to the development of safe and effective modern surgery are cocaine and tubocurarine. Cocaine was the first local anesthetic. Tubocurarine, the principal toxicant in tube curare (curare stored and shipped in bamboo), was the first muscle relaxant used in general anesthesia.

Long before European and North American scientists and physicians co-opted cocaine and tubocurarine, the power of these toxins was discovered by various Indigenous cultures throughout the Andes Mountains and

Amazon River basin. Like many of the toxins I have discussed, cocaine and tubocurarine can cure or kill, depending on the circumstances.

Cocaine is produced by South American coca plants in the genus *Erythroxylum*. DNA evidence shows that all cultivated coca varieties were originally domesticated by Indigenous peoples independently at two to three different times from the widespread wild progenitor *E. gracilipes*. Each variety falls into the species *E. coca* (Amazonian coca and Huánuco or Bolivian coca) or *E. novogranatense* (Colombian coca and Trujillo coca).

The question is why coca plants would bother to make cocaine in the first place. In 1993, James Nathanson, the same researcher whose studies proved that caffeine was a natural insecticide that protected tea and coffee plants, led the team that found the answer.

Nathanson and his team sprayed the natural host plant for the tobacco hornworm caterpillar with increasing concentrations of cocaine. At the concentrations found in natural coca leaves, caterpillar feeding was inhibited by up to 68 to 88 percent, and when the researchers applied a concentration that mimicked the higher cocaine levels found in newly opened coca leaves, feeding was inhibited by 95 percent. In both experiments, all the caterpillars eventually died. Like caffeine, cocaine is an incredibly effective insecticide, but it acts through a different mechanism.

The researchers found that cocaine kills insects by inhibiting nerve cell uptake of the neurotransmitter octopamine, which acts like norepinephrine in humans. The extra octopamine now available to bind to its receptor causes faster firing of neurons. The rapid firing leads to hyperactivity, including tremors, and causes the insect to quickly crawl off the leaves and stop feeding. All these behaviors benefit the plant because the caterpillar removes less leaf material.

Nathanson concluded that because cocaine inhibits feeding in insects, the alkaloid seems to serve as a chemical defense against attack by insects. Our use and abuse of cocaine is an "unrelated side effect" (Nathanson's words) of the fact that cocaine will inhibit the uptake of many neurotransmitters, including dopamine, serotonin, and norepinephrine, in mammalian brains like ours.

Cocaine also inhibits the hatching of the tobacco hornworm moth's eggs. More recent studies on fruit flies and ants found cocaine to be highly aversive in taste tests at biologically relevant concentrations in the food. Cocaine also shortened the life span of fruit flies and inhibited the development of their eggs. These results are entirely consistent with the insecticide hypothesis to explain why coca plants make cocaine.

Not surprisingly, the caterpillars of one moth specialize in eating leaves of the coca plant. The coca tussock moth *Eloria noyesi,* or the *malumbia,* as it is known in Colombia, somehow avoids the toxic effects of cocaine. Its caterpillars may even use the cocaine from the plant to protect themselves from predators by excreting it from either end when harassed. The caterpillars also seem to sequester the chemical during metamorphosis — cocaine is found in the adult moths, too.

This tiny moth has even attracted the attention of politicians. As entomologist May Berenbaum has pointed out, the George H. W. Bush administration spent $6.5 million to determine whether these caterpillars could be mass reared and then dropped from airplanes onto coca plantations in Colombia as part of the war on drugs. This same idea was resurrected in 2005 and 2015 by Colombian politicians as an alternative to the use of herbicides like glyphosate to destroy the coca plants.

The US government's willingness to spend such an inordinate amount of money on the coca tussock moth is a testament to how addictive cocaine is in humans. Why is cocaine so addictive? At the heart of the matter is the ability of cocaine to inhibit the recycling of neurotransmitters in the brain.

Cocaine is different from many of the widely used psychoactive drugs like opioids because it doesn't mimic the structure of a neurotransmitter (see the appendix online). Instead, it binds to three transporter proteins that sit in the cell membranes of neurons. These transporters bring dopamine, serotonin, and norepinephrine molecules back into the transmitting cell, the cell that releases these neurotransmitters.

Normally, the neurotransmitters released by the transmitting cell travel across the synaptic cleft, the tiny space between the transmitting cell and the receiving cell. Then the neurotransmitters activate the receptor in the receiving cell, and it fires. When that's done, the same neurotransmitters make their way back to the transmitting cell that first released them and are transported into the cell for the next round.

When cocaine binds to the dopamine transporter protein in the transmitting cell, the cell cannot remove dopamine from the synaptic cleft. The accumulation of dopamine drives prolonged firing of dopamine-dependent signals and, along with similar effects on the norepinephrine and serotonin transporters, produces a euphoric high.

Cocaine exerts its strongest effects in the nucleus accumbens, the part of the dopamine-driven mesolimbic reward system of the brain, and drives dependency, spurring cocaine use disorder. The situation is the opposite in insects because cocaine influences the reuptake of the neurotransmitter octopamine, which in turn causes insects to be *repelled by*, not attracted to, cocaine. This difference in neurophysiology explains why insects are repelled but humans are quickly captivated by cocaine and want to use it again and again.

As with caffeine and nicotine, which are also potent insecticides, humans co-opted cocaine for its role as a psychoactive drug millions of years after it first evolved. Cocaine's effect on us is simply a serendipitous, unintended consequence of the evolutionary battle between coca plants and the insects that eat them, and its effects on us do not explain why this substance arose.

The use of refined cocaine is entirely different from the practice of coca leaf chewing in South America. Coca leaves are among the most ancient psychoactive stimulants used by humans, and the practice of chewing them is essential to the modern identities and cultures of Indigenous Andean and Amazonian peoples.

The oldest evidence for human use of coca is more than ten thousand years old and comes from the western slope of the Andes in northwestern Peru. Chewed coca leaves from this site were radiocarbon-dated to 8000 BCE. In addition to the leaves, evidence of processed lime made from ashes of plants was found at this preceramic site.

Aymara and Quechua people still prepare lime the same way it was prepared in these ancient sites. They add it to the coca leaves to form a quid. The alkali properties of lime facilitate the release of cocaine from the leaves. To describe the use of coca leaves as chewing isn't quite right, because the leaf juices and saliva are sucked out of the quid and swallowed. Another traditional method involves grinding toasted coca leaves with the ashes of different plant species in the genus *Cecropia*.

Either method of preparation can produce a dose of cocaine in the range of 15 to 50 milligrams and a cocaine blood level of around 150 nanograms per milliliter. This level is far lower than that achieved by refined-cocaine users, who may consume up to 10 grams per day.

Today, there are roughly six to eight million *coqueros*, or coca leaf chewers, in South America, around the same number as there are regular refined-cocaine users in the United States alone. *Coquéo* (or *cocaísmo*), the chewing of coca leaves, helps stave off hunger, reduce fatigue, and combat the symptoms of altitude sickness, a constant threat on the Andean plateau. Coca leaf chewing and offerings are important components of ceremony and spiritual practice, contemporarily and historically. The practice is a daily ritual for millions of people, and it played a role in the rise and fall of the Incan civilization.

The Incans built central storage facilities for coca leaves, which were linked to the distribution of leaves to laborers. Early in the Spanish conquest of the Incans, attempts to eradicate coca use were soon replaced by an exploitative approach that included redistribution (selling) of coca leaves to agricultural workers who wanted to use them as their ancestors did. The stigma associated with coca chewing arising from racism may be one reason coca chewing never caught on in Europe.

Cocaine as a drug was therefore slow to enter the European and North American pharmacopoeia. John Stith Pemberton, a colonel in the Confederate States Army, was wounded in the Civil War and was treated with morphine. When he became addicted, he set out in search of a cure. Pemberton turned to coca and found it effective. In 1885, he formulated Pemberton's French Wine Coca, "the great remedy for all nervous system disorders." The tonic was inspired by *Vin Mariani*, a mixture of Bordeaux and coca leaf extract first concocted in Corsica by Angelo Mariani. However, a local alcohol prohibition ordinance in the state of Georgia forced

Pemberton to drop the wine from his brew, and the first formulation for Coca-Cola followed a year later.

The initial recipe for this patent medicine included coca leaf extract and caffeine extracted from cola (or kola) tree nuts from Africa, hence the name Coca-Cola. The product went on sale on May 8, 1886, at the soda fountains of Jacob's Pharmacy in Atlanta, where it was marketed to a white clientele as a temperance drink and "an esteemed brain tonic and intellectual beverage."

Racism motivated attitudes toward its use in Georgia, but in this case, the drug's use was exclusive to the oppressors, not the oppressed. The opposite situation held in the Spanish-Indigenous dynamic in South America.

In response to a public backlash against the drug, cocaine was removed from the ingredient list after 1903. Nevertheless, some news articles still report that "de-cocainized" Trujillo coca leaves are still being used to flavor Coca-Cola, possibly as the secret ingredient mysteriously called Merchandise No. 5.

If I had to guess, I'd say that modern-day Coke is also flavored by methyl salicylate, the main ingredient in the refreshing oil of wintergreen I mentioned before. But this guess is pure speculation. Nonetheless, methyl salicylate is a major component of the coca leaf's essential oil.

At least until the late twentieth century, there were news reports of a unique arrangement between the Coca-Cola Company and the Drug Enforcement Agency to obtain the de-cocainized coca leaves for flavoring. According to these stories, roughly 100 metric tons of coca leaves were legally imported into the United States annually by the Stepan Company in New Jersey. Because cocaine is still used in some surgeries, a supply is needed. After the secret ingredient was allegedly removed by Coca-Cola, the cocaine-containing extract was reportedly sent to Mallinckrodt Pharmaceuticals in St. Louis, where cocaine hydrochloride was refined. The veracity of these reports is unclear. Nor can the accuracy of these reports from decades ago, when the news stories first surfaced, be verified. However, by 2023, a simple internet search showed that Mallinckrodt was selling cocaine hydrochloride as a compounding drug, which is used in local anesthesia by oral surgeons and ear, nose, and throat specialists. Incidentally, Mallinckrodt was also one of the principal producers of oxycodone at the height of the prescription opioid epidemic.

In *Über Coca*, published in 1884, Sigmund Freud infamously extolled the virtues of cocaine as a stimulant but mentioned the anesthetic properties just once:

> Cocaine and its salts have a marked anesthetizing effect when brought in contact with the skin and mucous membrane in concentrated solution; this property suggests its occasional use as a local anesthetic, especially in connection with affections of the mucous membrane.

Freud used cocaine compulsively for over a decade, but the irony is that he began to study its effect (on himself) because of a friend who was addicted to morphine. Freud thought that cocaine might serve as a cure for morphine addiction. Instead, he used it to treat his own mental illnesses, including depression and AUD.

The same year that *Über Coca* was published, Carl Koller, one of Freud's medical colleagues in Vienna, discovered that cocaine could be used in eye surgery, specifically cataract removal, an otherwise incredibly painful procedure. Cocaine's use as a local anesthetic led to the development of other local anesthetics, including procaine (Novocain) and lidocaine (Xylocaine).

Although cocaine was a breakthrough in local anesthesia, curare was transformative for general anesthesia. Before physicians Harold Griffith and Enid Johnson introduced the alkaloid tubocurarine (the pharmacological agent in tube curare) into clinical practice in Montreal in 1942, there was no method of relaxing the body's muscles during surgery. This limitation made major surgery dangerous for patients and difficult for surgeons.

If there was ever a toxin that embodied the twin sides of natural toxins as both poison and cure, it is curare. The term *curare* describes many concoctions invented by diverse Indigenous cultures across the Amazon basin of South America using toxins from plants and other organisms. In that context, curare was mainly used to tip blow darts for hunting animals. This use explains how the toxin became known as "flying death" among colonizers and settlers.

In 1800, the German polymath Alexander von Humboldt and his companion Aimé Bonpland were the first Europeans to witness the preparation of curare in South America. Their description of the scene and what the "chemist of the place" intimated to them is notable:

> He was the chemist of the place. We found at his dwelling large earthen pots for boiling the vegetable juice, shallower vessels to favour the evaporation by a larger surface, and leaves of the plantain-tree rolled up in the shape of our filters, and used to filtrate the liquids, more or less loaded with fibrous matter. The greatest order and neatness prevailed in this hut, which was transformed into a chemical laboratory. The old Indian was known throughout the mission by the name of the poison-master (amo del curare). He had that self-sufficient air and tone of pedantry of which the pharmacopolists of Europe were formerly accused. "I know," said he, "that the whites have the secret of making soap, and manufacturing that black powder which has the defect of making a noise when used in killing animals. The curare, which we prepare from father to son, is superior to anything you can make down yonder (beyond sea). It is the juice of an herb which kills silently, without any one knowing whence the stroke comes."

After reading that passage, we can easily understand why curare is such a valuable product in the Indigenous communities of the Amazon basin and why the oral histories transmitted from generation to generation in Indigenous cultures have been the source of the many drugs we now have. It is also an example of how curare, a brilliant Indigenous technology, was superior to the guns used by European colonizers when it came to hunting prey in the rain forest.

Curare is typically prepared from the bark of either *Chondrodrendron* and related vines in the Menispermaceae family or *Strychnos* trees in the Loganiaceae family. Like cobratoxin from some cobras, tubocurarine blocks the nicotinic acetylcholine receptors, preventing the neurotransmitter acetylcholine from binding. When these receptors are blocked, the muscles that require an input from the nervous system to contract are unable to do so. That's what makes tubocurarine the perfect poison for hunting. Arboreal animals hit with a tubocurarine-tipped dart simply stop breathing and fall out of the trees after a few minutes. Moreover, the toxin is not poisonous when ingested, so there is no risk in eating the meat of these animals.

In 1935, British researcher Harold King obtained from the British Museum an old specimen of tube curare and isolated the alkaloid tubocurarine from it. Meanwhile, word spread that tube curare might be a useful drug. It was American adventurer Richard Gill, who had recurring muscle spasms so severe he couldn't walk, who learned from Indigenous peoples in Ecuador the secrets of tube curare preparation.

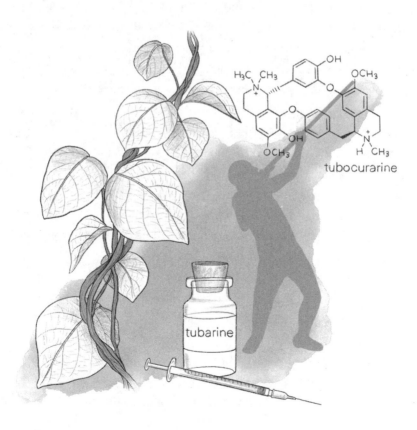

tubocurarine

tubarine

In 1938, Gill shipped a large amount of tube curare, along with specimens of the plants used to make it, back to the United States. Eventually, the pharmaceutical company Squibb began to sell a crude extract called intocostrin. But it wasn't until 1942, when Griffith and Johnson treated a patient with intocostrin and reported their success in the medical literature, that it gained traction.

In parallel, because of King's isolation of tubocurarine, work was already under way in the United Kingdom to develop a drug from the alkaloid itself. Eventually, Burroughs Wellcome introduced Tubarine, the brand name for tubocurarine.

By 1941, thirty thousand patients had been anesthetized with the revolutionary tubocurarine. Still, there were risks. Although the drug was almost miraculous as a muscle relaxant during surgery, tubocurarine also caused low blood pressure and severe allergic reactions. In 1946 and 1947, two physicians, one in Britain and one in Utah, did the unimaginable — they independently "curarized" themselves in a laboratory setting so that their colleagues could study how tubocurarine worked.

Such self-experimentation is not only ethically problematic but also physically dangerous; it would never be allowed to proceed today. Still, the reports these self-experimenters published in the scientific literature included crucial recommendations on dosing.

The experiences these two men endured while they were totally paralyzed is excruciating to read about. They were completely aware of what was going on around them but suffered what you could only imagine waterboarding might feel like — choking on their own saliva, unable to communicate that they were not getting the ventilation they needed. Thankfully, both men recovered completely.

Eventually, new, safer muscle relaxers were developed, but tubocurarine paved the way. It also sparked research on breathing physiology — research that helped lead to the development of lifesaving medical devices like the ventilator.

The connection between tubocurarine and the venom of the black widow spider is worth mentioning. Although only rarely lethal, the bite of this spider is rightly feared because of the pain caused by the release of acetylcholine its venom triggers. Because tubocurarine blocks the nicotinic acetylcholine receptor, it was used briefly in the 1950s — before an antivenin became available — to reverse the spider's toxic effects.

For scopolamine, cocaine, and curare, Indigenous knowledge, curated over tens of thousands of years, led to modern medical breakthroughs that have improved and extended countless human lives through anesthesia. In each case, the poison is the cure.

The Indigenous peoples who discovered the power of cocaine and curare never received compensation by the pharmaceutical companies that profited from them. It is no wonder that many countries in Latin America and elsewhere in the global tropics now have biopiracy laws that strictly regulate the export of natural products.

The use of these ancient drugs in modern medicine has waned, but another class of alkaloids has certainly not: the opioids. But unlike the poison-as-cure trajectory of the three toxins described in this chapter, the opioids follow a different pattern: the cure as poison. The next chapter describes these important alkaloids, their impact as medicines and addictive drugs, and, as we will shortly see, their direct and indirect roles in shaping world geopolitics.

9.

Opioid Overlords

On and on they walked, and it seemed that the great carpet of deadly flowers that surrounded them would never
end. They followed the bend of the river, and at last came
upon their friend the Lion, lying fast asleep among the
poppies. The flowers had been too strong for the huge
beast and he had given up at last... "We can do nothing
for him," said the Tin Woodman, sadly; "for he is much
too heavy to lift. We must leave him here to sleep on forever, and perhaps he will dream that he has found courage
at last."

—L. FRANK BAUM, *THE WONDERFUL WIZARD OF OZ*

Frankincense and Myrrh

During fieldwork in the Galápagos Islands, I frequently used the shade of
the palo santo, or holy stick tree, as a respite from the punishing equatorial
sun. I took refuge under its fragrant branches while banding and then
releasing the hawks I studied.

The silvery bark of the palo santo trees contrasted with the lava strewn
in red and black across the flanks of cinder cones. Although beautiful to
behold, it was the fragrance and ensuing sense of calmness of these trees
that is seared in my mind.

The aromatic resins of the palo santo and relatives in the Burseraceae family, which includes frankincense and myrrh, have been used since time immemorial in anointings, for embalming, as incense, in ritual cleansing, and as medicines. Palo santo was used by the Incans, and frankincense and myrrh by Africans, Europeans, peoples throughout the Middle East, and in Ayurvedic, Chinese, and Perso-Arabic traditional medicine.

Frankincense and myrrh have long been used as mind-altering drugs. The Gospel of St. Mark contends that wine containing myrrh was given to Jesus Christ before his Crucifixion, perhaps in an attempt to dull the pain.

New research helps explain why these resins have been so coveted and widely used to treat our ailing bodies and minds for so long. Extracts from frankincense provide pain relief and produce sedative effects when injected into lab rats.

One of the principal terpenoids in frankincense, incensole acetate, reduces anxiety and depression symptoms in lab mice. It seems to do this by binding to the TRPV3 receptors in the brain. Interestingly, other TRPV3 receptors are found in the skin and play a role in the sensation of warmth, both thermal and chemical. For example, carvacrol, thymol, eugenol, and other phenolic and terpenoid chemicals found in the essential oils of oregano, thyme, and clove activate these TRPV3 receptors when the oils are rubbed into the skin. The result is a wonderful feeling of warmth. Together, these findings give us one biological explanation for the ancient and widespread use of frankincense in spiritual rituals and medicine.

Myrrh is even more remarkable. In controlled experiments, laboratory mice were fed ground myrrh or its main terpenoid, furanoeudesma-1,3-diene. Each treatment conferred higher pain tolerance than that of control mice only given saline. Incredibly, when these mice were given the opioid receptor blocker naloxone, the analgesic effects of myrrh disappeared.

Using radioisotope tracers, the researchers then confirmed what you may already suspect: the furanoeudesma-1,3-diene of myrrh seems to bind directly to opioid receptors in the brain and had pain-relieving effects comparable to morphine. The receptors are the same ones that opioids and our own endorphin peptides bind to.

Furanoeudesma-1,3-diene holds promise as a less addictive pain reliever than opioids. To be clear, this terpenoid is an opioid-like compound but not an actual opioid. In nature, opioids are alkaloids produced by the opium poppy.

Mesopotamians, Mayans, and Morphine

The usual, and incorrect, origin story for the opium poppy is that according to ancient Sumerian writings and the art of their Assyrian descendants, the Sumerians domesticated the plant, and then its use spread to the Egyptians. This tale was propagated through the writings of Anthony Neligan, physician to the British in Persia. He claimed that as early as 7000 BCE, the Sumerians first harnessed the poppy's powers.

However, the ancient Assyrian tablets that discuss the opium poppy were in fact nowhere near nine thousand years old. They came from the Royal Library of Ashurbanipal, dated to roughly seven *hundred* years before the Common Era, not seven thousand.

Other evidence offered under the Sumerian origin theory is from Assyrian cuneiform tablets that use the ideogram HUL-GIL. Assyriologist Raymond Dougherty interpreted HUL-GIL as "joy plant," a term that, he tentatively suggested, may refer to the opium poppy.

Dougherty's take doesn't hold up, either. Another Assyriologist, Erica Reiner, reexamined the evidence and concluded, "No word either in Akkadian or Sumerian has been identified as opium poppy…As for the Sumerogram HUL-GIL meaning 'joy plant,' I have to say that this is completely erroneous. Indeed, the first sign is…'cucumber.'"

As for the depictions of opium poppies in the art of the Assyrians and Egyptians, the popular eye-of-the-beholder idea comes into play. Bas-reliefs and a statue from the Altar of Hapi, god of the Nile, both widely interpreted to depict opium poppies, can alternatively be interpreted as pomegranates and blue water lilies.

The first people to cultivate the opium poppy, *Papaver somniferum* subspecies *somniferum*, were Neolithic farmers in the Mediterranean region of Europe, around 5600 BCE. The wild progenitor of the opium poppy is

P. somniferum subspecies *setigerum,* which still grows wild in the central and western Mediterranean region. The oldest radiocarbon-dated specimens are from La Marmotta, a site near the Italian town of Anguillara Sabazia, about twenty miles northwest of one of my favorite cities, Rome.

By 5300 BCE, opium poppies had reached northwest Europe, and three hundred years later, they were cultivated in the western Alps. The opium poppy may well be the only seed crop domesticated in Europe. Poppy seeds were eaten and turned into valuable oil, and, of course, the morphinan alkaloids (chemical precursors to opioids) in latex were used as medicinals.

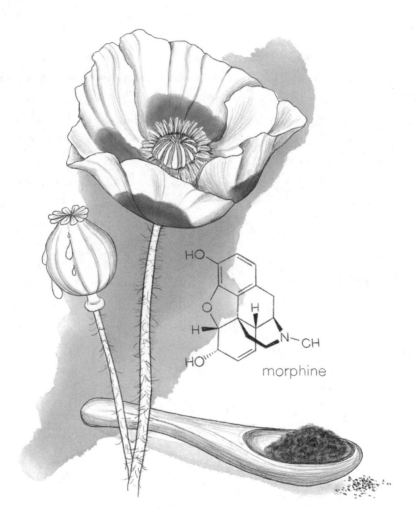

morphine

The evidence suggests that people were using the opium poppy in Mesopotamia by the middle of the third millennium BCE, in the late Early Bronze Age. For example, a remarkable kitchen with eight hearths, dated to 2450 to 2300 BCE, was discovered in the Royal Palace G at Ebla, Syria. This was not a normal kitchen—or at least not a kitchen we'd recognize. Instead, it was used to make and store extracts from medicinal plants, including, apparently, opium poppies.

Base-ring juglets made in Cyprus were traded throughout the Eastern Mediterranean in the Late Bronze Age (1650–1350 BCE). Because of their unmistakable inverted poppy-like shape, in 1962, archaeologist Robert Merrillees proposed—controversially—that these vessels transported opium.

In 2018, scientists had the technology to test his bold idea. They drilled into a Cypriot juglet at the British Museum, sampled residues within it, and subjected them to sophisticated chemical analysis. The process was similar to how chamazulene was recovered from Neanderthal teeth. The juglet held traces of both thebaine and papaverine, two of the more stable morphinan alkaloids produced by the opium poppy. Merrillees was vindicated.

When opium poppy supplies were running low in France during World War II, scientists searched for a replacement narcotic. Although the term *narcotic* is mostly used to describe opioids today, a broader definition truer to the Greek root *narkoun,* which means "to make numb," is this: "having the effect of relieving pain and inducing drowsiness, stupor, or insensibility."

The French turned to the tubers of the local white water lily, *Nymphaea alba.* After experimenting on small animals, they found that "in all instances, there was elicited a narcosis that terminated in somnolence." In other words, the animals fell into a narcotic-induced sleep.

Eventually, the drugs derived from these water lilies became important medications. Tracing their origins requires stepping back thousands of years, when they were used by the most sophisticated two societies in the world, but on opposite sides of the planet—Mesoamerica and Egypt.

Blue water lilies (*N. noucahli* var. *caerulea*) played a central role in the theology of the ancient Egyptian priestly caste; the plants occur throughout the Nile delta. The blue water lily appears in several beautiful ancient Egyptian creation stories. From the lily emerged the deities of the pantheon. In one story that emerged from the city of Heliopolis, the world sprang from Nun, a "watery chaos" that gave rise to the blue water lily, and from it arose Atum, the sun god. Apropos of this god, the flowers of water lilies actually close at night and open at dawn, coinciding with the rising sun.

In another creation story from the ancient Egyptian city of Memphis, Osiris was slain and tossed into the Nile and was later resurrected as a blue water lily. He fathered Horus, whose four sons emerged together from a flower of a blue water lily.

In chapter 81 of the Papyrus of Ani, or *Book of the Dead*, Nefertem emerges from a blue water lily as the world is created. Tutankhamun was depicted as a child in the striking painted wooden bust known as the Head of Nefertem. Young King Tut's neck emerges from a blue water lily painted in a teal polychrome.

In complex ritualistic fashion, the priests used the lily as a narcotic to reenact these successions of deities as they were believed to have appeared across the ages. The priests appeared to have believed they were transitioning from one deity to the next after they had taken the blue water lily concoction.

A related water lily from Mesoamerica (*N. ampla*) was also exploited by Mayan priests in astonishingly similar ways halfway around the world. In both civilizations, artisans carefully incorporated stunning depictions of water lilies on frescoes and on many other artifacts. In each case, shamans also used them to reenact the transformations of deities, from one animal or human form to another.

So similar were the ancient Egyptian and Mayan cults of the water lilies that ethnobotanist William Emboden concluded that "the two water lilies have common belief systems associated with them that are patterned and predictable especially when one considers the drug properties of such plants." The remarkable and undeniable implication is that the use of water lilies as a psychoactive plant evolved independently in two of the most highly advanced societies in the world at that time.

We now know that both water lily species contain aporphine alkaloids that, depending on their form, can either block or activate the D_1 and D_2 dopamine receptors in our brains. Aporphine, one such alkaloid, blocks the brain's dopamine receptors and is converted to apomorphine by the body. Apomorphine then activates the dopamine receptors. Today, apomorphine is sold as the drug Apokyn, which is used to treat Parkinson's disease and erectile dysfunction.

Another water lily alkaloid, nuciferine, is a dopamine blocker, but its breakdown product atherosperminine has the opposite effect and activates the dopamine receptors.

Given the opposing effects on the dopamine receptors as the drugs from the water lilies are broken down in the body, consumption of these alkaloids by the ancient Egyptian and Mayan cultures would have produced a range of mental states that depended on the amount taken and how quickly the drugs were metabolized in the body. Such opposing effects were exploited in the priestly rituals of both cultures to reenact the successions of their deities.

aporphine

So, now it makes sense why waterlilies, and not opium poppies, were probably depicted on the ancient Egyptian Altar of Hapi. Eventually, however, opium did make its way into ancient Egyptian culture but only after the poppies were domesticated in what is now Italy.

Europe was the site of domestication of the poppy and was where, in 1805, Prussian pharmacist Friedrich Wilhelm Sertürner first isolated morphine in pure form. He called it morphium, after Morpheus, one of the thousands of sons of Somnus, the god of sleep in Ovid's *Metamorphoses*. Varieties of the opium poppy produce variable amounts of the benzylisoquinoline, or morphinan alkaloids, that include codeine, morphine, and thebaine.

After boiling opium in water, Sertürner used ammonia to obtain morphine crystals. Then he did something unthinkable. He persuaded three teenage boys, "none older than seventeen," to take the drug with him to prove that it acted in the same fashion as opium. They all took the drug by dissolving half a grain of the crystalline morphine in alcohol and water and continuing to take this dose in rapid succession until, as he wrote, "the outcome with the three young men was decidedly rapid and extreme."

By 1827, Merck began morphine production in Germany, and the drug was used intensively to treat wounded soldiers during World War I. Morphine is still one of the most widely used and effective ways to treat pain (as is codeine).

The alkaloid thebaine, from the opium poppy, is less well known because it has different pharmacological properties from those of morphine. Thebaine works by blocking opioids and endorphins (the endogenous peptides we produce to block pain) from binding to the opioid receptors in the brain instead of activating the receptors as morphine does. Because thebaine is a precursor to morphine itself, it can be converted into two of the principal drugs at the heart of the last wave of the opioid epidemic: hydrocodone and oxycodone.

The traditional method of harvesting opium required lacerating the developing capsule, which contains the seeds, and then allowing the latex to seep out and dry, the so-called opium stage. Later, the dried latex or raw opium must be scraped off for direct use or further processing, a labor-intensive process.

The modern method of harvesting and extracting alkaloids is far less labor-intensive. The entire above-ground biomass of dried, mature plants, known as poppy straw, can be harvested by machine, using a technique developed in 1930 by Hungarian pharmacist Janos Kabay.

In 2004, scientists reported a new opium poppy mutant produced through plant breeding methods. Called *top1,* short for *thebaine oripavine poppy 1,* the genetically engineered plant accumulates thebaine and the morphine precursor oripavine but not morphine itself. Since morphine is not required to produce hydrocodone and oxycodone — only thebaine is needed — *top1* poppies represented an opportunity for pharmaceutical companies. That's because separating thebaine from morphine in opium poppies is difficult, and with *top1* poppies, pharmaceutical companies didn't need to do this separation step.

As a bonus, *top1* poppies would not be useful to the illicit drug trade, because heroin is made from morphine, not thebaine. The poppy mutant was a win-win, except for the millions of people who wound up addicted to the plentiful opioids derived from *top1* poppies.

Thebaine is also used to make naloxone (sold as Narcan), the antidote to acute opioid overdoses, and buprenorphine, a treatment for opioid withdrawal symptoms. Finally, thebaine can be converted into naltrexone, which is used to treat drug cravings in people with opioid and alcohol use disorders.

As I described earlier in remembering my father, I was well aware of these disorders. The two prescription pill containers my brother had found in our dad's fifth-wheel trailer had the word HYDROCODONE blazoned across the labels. The word gave me a sinking feeling when I read it. Strangely, however, the pills had been prescribed to his brother, not him. His brother had been dead for several years, although my father initially had moved in with him after running away and before permanently moving to Texas.

I wondered why he had held on to his brother's hydrocodone pills for so long. Was my father addicted to opioids in addition to alcohol? Did he die of an opioid overdose?

On June 3, 2018, we received the medical examiner's report from the justice of the peace. There were no big surprises as I began to read it, just a reckoning with what AUD does to the body. The examiner's first finding was that my father had cirrhosis of the liver and severe coronary artery disease.

I had somewhat expected this. His decades of heavy drinking caused the cirrhosis, increased his risk of developing atherosclerosis of the coronary arteries, and increased his risk of heart attack by 40 percent. His blood alcohol level at the time of the autopsy was around 0.1 percent, just above the legal limit when driving in Texas.

My initially sanguine reaction to the situation suddenly changed as I read the medical examiner's next finding: "Aorta blood drug screen positive for methamphetamine (ELISA), amphetamine (ELISA), fentanyl (ELISA)." ELISA is a testing method, enzyme-linked immunosorbent assay that uses antibodies to detect the presence of certain chemicals in a blood sample.

Pepper, Pork, and Piperidine

My father and my maternal grandmother had much in common, although of course they were not related biologically. They were both outwardly warm people whose matching green eyes and quiet smiles could charm any room.

Her parents hailed from the British Isles, and his ancestors did too. My dad was her handyman. He was reliable and eager to please his mother-in-law the matriarch, their dynamic a template for how people should treat one another. I watched him climb ladders to change her light bulbs, fix her broken vacuum cleaner, and tighten the screws in her old chairs — all under her watchful eye.

Their fondness for each other was subtended by a shared understanding that each would need repeated slugs of ethanol before dinner — whiskey gingers on the rocks for her and cold beer, of course, for him. Their $GABA_A$ receptors fired away as the alcohol did its thing, dampening their worries, numbing their pain, and transforming them into different people.

Perhaps the strangest trait they shared was that they both heavily over-flavored their food with black pepper. I mean *heavily*. From an early age, I had mixed feelings about the dinners they prepared. My trepidation was not due to the food itself, which was delicious. It was the copious black pepper that they endlessly cracked over it.

For my grandmother, the black pepper vector was usually a pork roast that the adults raved about. For my father, it was pot roast that I only began to appreciate in my late teens.

Black peppercorns are produced by the *Piper nigrum* plant. Black gold, as it was known, was a primary motivator for the early spice trade and all the geopolitical consequences that flowed from it. The spice tingles the tongue because of the alkaloid piperine.

Piperine binds to TRPV1, the same receptor activated by capsaicin, eugenol, gingerol, and so many other chemicals in spices. When triggered, the receptor activates the pain-sensing nerves in our mouths. As we up the dose, the overuse of piperine eventually desensitizes these same nerves.

The overstimulation of pain receptors like TRPV1 is followed by an inability to feel pain. This loss of pain perception stems from the activation of both the dopamine and endogenous opioid pain-relieving systems and the mesolimbic reward center. Overuse of these spices might trigger a pain-inhibits-pain feedback loop.

As you might have guessed, the piperine in black pepper is a potent defense against attack by most herbivores. But some insect species are well adapted to it. The pollu beetle is one. A native pest in India, its grubs feed on the piperine-rich berries that contain the black peppercorns.

The technical name of piperine is 1-peperoylpiperidine. Piperidine forms the backbone of piperine and can be artificially synthesized by reacting piperine with nitric acid. In this way, piperidine became the basis for the synthesis of pethidine in 1938. Also known as meperidine, pethidine would later be sold as the prescription pain reliever Demerol.

Pethidine soon eclipsed morphine as the prescription analgesic of choice for acute and chronic pain in the mid-twentieth century. Yet, like morphine, pethidine had undesirable side effects: neither drug could penetrate the blood-brain barrier well. A drug must be able to penetrate this barrier to reach the brain. Pethedine's therapeutic effect had a slow onset, and the difference between toxic and therapeutic doses was narrow.

Still, pethidine's structure was less complex than morphine's and easier to manipulate in the lab. So in 1953, Paul Janssen founded the Belgian company Janssen Pharmaceutica and began to tinker with pethidine as a starting point to synthesize more effective and, he hoped, safer opioids.

Janssen hypothesized that the piperidine backbone of morphine and pethidine was what allowed them to bind to opioid receptors in humans. He was right, to a fault. His first commercial success was phenoperidine, which was synthesized in 1957. Phenoperidine was followed by fentanyl in December 1960. A piperidine ring lies at the heart of Janssen's fentanyl molecule.

We've encountered piperidine alkaloids, as they are collectively known, several times in the book. They've been mentioned in discussions of black pepper, the eastern white pine needles on my boutonniere, and the medical examiner's report on my father's autopsy. And piperidine alkaloids also lie at the heart of the fentanyl crisis.

Fentanyl, which binds to the brain's opioid receptors, is one hundred to three hundred times more potent than morphine. When modified slightly into the opioid carfentanil, it is ten thousand times more potent than morphine. Fentanyl is one of the most widely used treatments for pain relief in the world. In the United States, more than 1.4 million fentanyl prescriptions were written in 2019.

In 2017, the year my father died, US physicians wrote nearly 192 million prescriptions for opioids—59 prescriptions per 100 people. High as that rate may seem, the number of prescriptions has come down since their 2010 peak of 81.2 per 100 people.

Nonetheless, the rate in the Texas county where my dad lived was 97.4 prescriptions per 100 people in 2017, just shy of one prescription per person. Many were legitimate prescriptions for surgeries and chronic pain management. Many were not. The amount of illicit fentanyl available in the United States is staggering. And carfentanil, the new and far more powerful piperidine-based opioid, is also emerging as a threat.

There have been three distinct waves of the opioid overdose epidemic.

The first began with the abuse of prescription semisynthetic opioid pills like hydrocodone and oxycodone. When those became less available, the second wave began as users turned to heroin. The third and current wave is fentanyl abuse, and it is unlikely to be the last.

There were 80,926 deaths from opioids in the United States in 2021, the most ever recorded. In the past twenty years, opioids have taken more than 500,000 lives. One in three Americans knows somebody who has had an opioid use disorder.

Despite these figures, I was surprised that the ELISA test on the medical examiner's report was positive for amphetamines and fentanyl. I kept reading, and the next finding confused me even more: "Aorta blood (LCMS) negative for methamphetamine, amphetamine, fentanyl." The LCMS (liquid chromatography mass spectroscopy) test is far more accurate than the ELISA and is similar to the test done on the Neanderthal's teeth. The medical examiner had used it to double-check the ELISA results. I'm glad they did.

The ELISA results were false positives. In the end, there was no link between the hydrocodone pill containers that my brother found, the three drugs initially detected, and my father's death. The medical examiner didn't list the drugs as causes of death, because they weren't really there — except for the ethanol.

The official cause of death was cirrhosis of the liver and coronary artery disease — each undeniably caused by his AUD. His liver and heart eventually failed, owing to the roughly twenty-five hundred beers he consumed each year for fifty-plus years. So it was complications from his AUD that took him down, not amphetamines or opioids. It was a strange relief to discover that it was the devil we knew, not the one we didn't, that killed him.

I still wanted to know why he used alcohol to take away the pain and then, ultimately, couldn't stop using it. The familial dysfunction, the sexual abuse, the physical pain, and the family history of AUD — all these things contributed to his susceptibility. But ever the biologist, I wanted more concrete answers. So I dug. My curiosity led me to understand the important links between the brain's homegrown opioids — the endorphins — and AUD.

Endorphins, Heroin Homebrew, and Alcohol Use Disorder

When plant-insect interactions really began to consume me toward the end of my PhD research, the change in direction from my original focus was unsettling. I was well on my way to becoming an expert in birds and their parasites, not on plants and herbivores.

Yet, switching fields comes naturally to me because it emerges from serial obsessions I simply can't control; the desire to switch washed over me with an intensity that lit up what seemed like every neuron in my brain. The feeling was there when I opened my eyes in the morning and just before I closed them at night, much as all my other obsessions flowing from nature have serially waxed and waned from my earliest days. I couldn't stop my craving to know more, even though I wanted that desire to go away.

Each obsession was like an addiction to a different drug. I had to consume as much information as I possibly could, until the wee hours of the night, before collapsing in bed. Information gathering was my drug of choice, and the reward was simply knowing and sharing, to whomever would listen.

First, my poor mother lugged book after book home from the local library and attentively listened to each encyclopedic entry that needed to be released from my brain. Then I turned to nature itself as a teenager, observing and absorbing directly.

My life as a PhD student was enriched by the best botanists in the world. I lived in St. Louis with my first boyfriend, a botanist, whose apartment was five hundred feet from the Missouri Botanical Garden in one direction and its Monsanto Research Center, where the graduate students hung out in the other, with a legendary mom-and-pop doughnut shop in between.

Other circumstances were also guiding my new interest. The Donald Danforth Plant Sciences Research Center in St. Louis had just opened, and most of my mentors at UMSL studied plants and insects. A green gravity began pulling me in. But I was unaware of what was unfolding at the Danforth center just as I was dipping my toe into plant biology.

Some years before, in 1985, physiologist Avram Goldstein published a study with an odd and potentially off-putting title: "Morphine and Other Opiates from Beef Brain and Adrenal." You read that right—morphine in the brains and adrenal glands of cows. It sounds just like the small amounts of homemade salicylic acid, cardiac glycosides, and DMT in us, right?

That same year, morphine was found in toad, rat, and rabbit skin. Shortly thereafter, codeine and thebaine were found in the brains of some mammals too. Then researchers discovered that mammalian livers could convert the morphine precursor reticuline into the morphine precursor salutaridine, a critical step in morphine biosynthesis only known from the opium poppy.

If it were true that humans make the same morphinan alkaloids as those made by the opium poppy, it would shake up our basic understanding of opioid biology. Major controversy ensued over the next three decades. The morphine may have simply been an artifact, brought in through the diet from some unknown source of the alkaloid, and not actually made by our bodies. If mammals did make morphinan alkaloids, they evolved the ability to do so hundreds of millions of years *before* the opium poppy did.

We humans can use morphinan alkaloids as painkillers because these molecules bind to the receptors that normally bind to our endorphins. Yet the two sets of molecules—opioids and endorphins—are completely unrelated chemically. Endorphins are peptides made of chains of amino acids, whereas opioids are alkaloids made of piperidine molecules.

In 2004, while I was in the middle of my dissertation work, Danforth researchers led by biochemist Meinhart Zenk reported the results of an experiment confirming not only that some mammals make morphine but also that we humans do. Zenk and his team bathed human cell lines in an atmosphere of oxygen isotopes. Because these isotopes are slightly heavier in atomic mass than the more common isotopes, the researchers could trace the biochemical reactions as the cells used the isotopes to make various chemicals of life. They then bathed the cells in dilute isotopically labeled morphinan precursor chemicals.

These two isotope-labeling experiments allowed the Zenk team to use highly sensitive mass spectrometry to test if any morphinan alkaloids could be produced by the human cells. In the human body, the microorganisms that make up our microbiomes might produce morphinans. To rule out this possibility, the researchers kept the human cells in a germ-free environment during their experiment.

Morphine and several other morphinan alkaloids were produced by the human cells. Six years later, living, breathing mice, not just cell lines, were shown to produce morphine too. This discovery was in line with the finding that mice and humans secrete small amounts of morphine in their urine. Zenk's research also added to the (albeit controversial) data that had been accumulating since 1985 on animal synthesis of morphinan alkaloids.

Human cell lines and mice are one thing. The real question was whether people really make morphine, and if so, what it is doing in our bodies.

Bear with me as I take us through some arcane details of how Parkinson's disease progresses. An understanding of this disease will lead us to a startling hypothesis about homegrown morphine in humans.

The first big hint that human brains make morphine came indirectly on the heels of Oliver Sacks's 1969 breakthrough (but only temporary) treatment for parkinsonism caused by encephalitis lethargica. The treatment was the drug levodopa (or L-DOPA), and Sacks wrote about this disease and the treatment in *Awakenings,* the basis of the eponymous film.

In 1973, the same year that *Awakenings* was published, researchers discovered the alkaloid tetrahydropapaveroline (THP) in the brains of Parkinson's patients who were treated with L-DOPA. Note the root word *papaver* in the chemical name. As noted earlier, *Papaver* is the genus of the opium poppy, and the inclusion of the genus name in the chemical term for this alkaloid is no coincidence.

Subsequently, THP was found in the urine of both humans and rodents, along with morphine. In light of the available evidence, scientists concluded that the human brain produces THP from dopamine and then uses it to make morphine.

L-DOPA is given to people with Parkinson's to stimulate dopamine production, but apparently there is an unintended consequence: our brains can also make THP from L-DOPA. Patients given L-DOPA had elevated levels of THP, which can be synthesized from L-DOPA, as an alternative to dopamine.

Humans' apparent ability to make THP is notable. This alkaloid is also a precursor chemical made by opium poppies in the production of the morphinan alkaloids codeine, morphine, thebaine, and others.

Remarkably, both humans and opium poppies can make THP, and as we now know, both can make morphinan alkaloids like morphine too. What all this means is that the metabolic pathway—in other words, the chemical capacity—to produce morphinan alkaloids evolved at least twice independently, once in mammals like us and once in plants like the opium poppy.

There is more to this story than just an interesting observation of the repeated path that evolution can take. A link has been found between THP levels in the brain and AUD. THP levels are elevated after ethanol consumption in rats, and in both rats and monkeys whose brains were infused with THP, they voluntarily drank ethanol in excessive amounts to intoxication. When the alcohol was taken away, they exhibited withdrawal symptoms, as in humans with AUD, like my father.

Although the theory is far from confirmed, some scientists suggest that THP is associated with AUD symptoms because it interferes with the mesolimbic dopamine-based reward system of the brain—the same system that amphetamines, cocaine, and opioids hijack. The ethanol may generate a reward-based feedback loop, which is partly mediated by production of THP in the brain. THP levels are low in abstinent alcoholics, but the brain makes more THP when it senses ethanol.

To sum up all this, it is clear we make morphine by using THP in small amounts. And in higher amounts, as the L-DOPA and experimental rats and monkeys show us, THP is clearly toxic and triggers behaviors consistent with AUD. This pattern is seen when we compare people with AUD who are abstaining from drinking and who therefore have low THP levels with people with AUD who are drinking and who have high THP levels.

In addition to the endogenous endorphin peptides, mammals therefore seem to make morphinan alkaloids such as codeine and morphine. Mammals evolved more than a hundred million years *before* the opium poppy did. This means that mammals were making morphine well before plants evolved the capacity to do so. The list of chemicals that are used as toxins by some organisms and that are also produced by our brains is now long. Even when there isn't an exact match, the defensive chemicals from these other organisms often either mimic the chemical messengers our nervous systems use or disrupt their function (see the appendix online).

A Brain Is a Brain

Owing to its bitter, quinine-like taste, morphine is aversive to animals when it is included in their diet. So, from the opium poppy's point of view, as a first layer of defense, this chemical's distastefulness works as a defense against repeated attack by herbivores. Yes, morphine also has narcotic effects on animals if enough is consumed, so this itself may drive deterrence. When a poppy capsule is wounded, some of the morphine in the latex is converted into bismorphine. This molecule forms cross-linkages with pectin, a constituent of plant cell walls. The cross-linked bismorphine strengthens the pectin, which may make it more difficult for enemies to attack the plant. This function for bismorphine hasn't been proven, though.

Morphine may serve as a multipurpose defense, like the psilocybin found in magic mushrooms. Remember that psilocybin sits in wait in the mushroom and, when converted into psilocin, directly messes with the minds of animals. But it also serves as a backbone for the blue, tannin-like molecules that produce oxidative stress in the bodies of the attacking insects that ingest it. A toxic two-step may be part of the overall defense strategy found in both opium poppies and magic mushrooms.

In laboratory experiments, fruit flies tend to avoid solutions laced with morphine as much as they do caffeine. If these insects take another sip despite the bitter taste, an even greater aversion to morphine develops over the course of an experiment. If mammals return to take another taste, by contrast, a preference for diet laced with morphine can develop.

Rats preferred food laced with bitter morphine over food laced with bitter quinine. And just like humans, other mammals, like rats, that are socially isolated or are under other stresses are prone to increase their preferences for, and dose of, morphine. This initial aversion followed by reinforcement is called the *paradox of drug reward*.

These patterns suggest that instead of having a reinforcing effect like it can have in mammals, morphine has the opposite effect in insects, driving them away. Cocaine has a similar repellent effect in insects but can have an opposite, reinforcing effect in us. Other experiments suggest that these differences between mammals and insects may depend on the circumstances.

Just as there are specialized moth caterpillars that only eat coca leaves, some insects are somehow rewarded by morphine. This finding is curious because insects are not known to have opioid receptors. But when carpenter ants were initially offered a sugar solution with morphine, they eventually self-administered the morphine even in the absence of the sugar reward.

Nobody yet knows what exactly is going on in these ant brains, but when neurotransmitter levels in the brains of the morphine-fed ants were measured, dopamine levels were higher than those in control ants. Morphine's ability to increase dopamine in the human brain can partly explain why people so easily start to abuse this substance. So, even in the brains of ants, we see remarkable clues about why humans might be prone to abuse opioids.

Another hint came from experiments that actually injected morphine into insects when they were purposefully injured. When praying mantises were given an electrical shock and when crickets were placed in an extremely hot box, injections of morphine increased the pain tolerance of both insects. But when they were then given the opioid blocker naloxone, the drug reversed the effect of morphine just as it does in people. Curiously, even though current evidence indicates that insects lack the opioid receptors that we humans have, *somehow* the opioids delivered what might be a pain-relieving effect in the insects. We still don't know why.

So, with training, some insects, like ants, come to self-administer morphine when no nutritional reward is present. What's more, insects have higher pain tolerance when injected with opioids—consistent with the behaviors of mammals, which are prone to drug use disorders.

These conflicting responses from insects—learned avoidance by fruit flies and learned and possibly dopamine-dependent preference in ants—present a paradox. But we might resolve it if we look at their social lives. Fruit flies are not the most social of animals, a characteristic that also applies to most herbivorous insects. Consequently, their brains might not be as vulnerable to the effects of the opioids. But ants are among the most social of all insects, and in key ways, their complex societies, which require cooperation between individuals—literal hive minds—to function, mirror our own in fundamental ways. So their apparent vulnerability to opioids comports with their extremely social nature.

The reinforcement of helping behaviors is important in both ant and human societies; cooperation between individuals is the secret to their success and ours. Even if ants don't seem to have opioid receptors like ours, the administration of morphine, not just saline, produces more of the rewarding chemical dopamine, just as it does in our brains.

Dopamine levels matter because drugs like amphetamines, caffeine, ethanol, nicotine, cocaine, and opioids all elevate it. These drugs are also generally bitter, distasteful, and toxic to enemies of plants like herbivorous insects. Yet we (and sometimes animals trained to do so in the lab) go back for more despite an initial aversion, thanks to the increase in dopamine in just the right place in our brains—the mesolimbic reward system.

The question is why we started using these addictive toxins in the first place. We can get at a general answer, by looking closely at animals that, like many herbivorous insects, have also overcome their aversion and then even evolved to co-opt the toxins for their own devices.

In both cases, whether it's an animal that eats the toxin to protect itself from being eaten or it's a person who uses the toxin to prevent pain, the user walks a knife's edge. In the end, there is no such thing as a free lunch. To get the benefit, you pay the cost. For monarch butterflies, for example, the youngest among them—the neonate monarch caterpillars—can't handle the toxins found in milkweed plants as well as the adults can, and many of these small caterpillars perish. For humans, the young who have experienced abuse, trauma, and neglect in their childhoods are particularly vulnerable to drugs of abuse.

In many ways, of course, we *are* different from these specialized animals, which evolved hardwired behaviors to specialize on toxic diets and innately seek them out. On the other hand, when we endure stress, pain, or repeated exposure to a toxin, our bodies respond to these chemicals much like those specialized animals do.

Many toxins present great paradoxes. On the one hand, caffeine, cocaine, ephedrine, morphinans, nicotine, and ethanol protect plants and fungi from attack. On the other hand, these chemicals can reinforce our consumption of them to the point where we can't stop going back for more. Instead of keeping us at bay, the toxins can spur some of us to go to great lengths and to spare no expense to obtain them, once a dependency has developed.

That the drugs exist at all is because plants and fungi want to live too and use toxic and rewarding chemicals to ensure that they do. Drug use disorders may simply be an unintended consequence of the way our brains are wired.

Although we can learn about human susceptibility to drugs of abuse by studying how animals respond to them, no wild animals fall prey to drug use disorders. To further resolve this paradox of drug reward, we will next focus on early humans' relationship with nature's toxins. Almost all drugs of abuse, even the psychedelics, were first used by Indigenous peoples as food or medicines. To a large degree, we are what we eat.

10.

The Herbivore's Dilemma

Tell me what you eat, I shall tell you what you are.
—Jean Anthelme Brillat-Savarin, *The Physiology of Taste or Meditations on Transcendental Gastronomy*

You Are What You Eat

Each of us faces a daily conundrum of what to put in our bodies. Yet, it is the people in our lives, the cultures in which we are embedded, and where we live on the planet, that largely determine what we consume. Our choices also depend on some basic biological differences between us, including our age, pregnancy status, sex, and whether we carry certain genetic variants in our DNA. These cultural, familial, environmental, and biological circumstances work independently and together to influence what we choose to eat, drink, smoke, chew, snort, or inject.

The youngest among us are far more limited in the choices they have. Forced to simply reject or accept what is offered, an infant or a young child who cannot yet communicate well is unable to make special requests. In the end, to be nourished, they must eat and drink from a subset of the items we give to them. This arrangement has far-reaching consequences. To understand these consequences, we first need to focus on how the incidental consumption of nature's toxins by infants and young children dramatically shapes both the quality of their adult lives and the evolutionary trajectory of our species.

Anthropologist Fatimah Jackson has developed an interesting

hypothesis in this regard—one that concerns a relationship among human malaria, sickle cell disease, and cassava root. But before we delve into her hypothesis, we need to understand the evolutionary and genetic aspects of malaria and sickle cell disease.

Falciparum malaria is caused by the parasite *Plasmodium falciparum* and is a major killer of infants and young children in sub-Saharan Africa. In 2020, around 600,000 of the world's 627,000 deaths from malaria occurred in this region. Tragically, those below the age of five accounted for 80 percent of the deaths.

Sickle cell disease, meanwhile, is an inherited human disease that results from a genetic change in the hemoglobin gene. Our red blood cells use hemoglobin to transport oxygen from the lungs. A single mutation in the hemoglobin gene has spread in some populations of humans over the past several thousand years and causes a change in one amino acid of the hemoglobin molecule. This change has a big impact: normal disk-shaped red blood cells turn into rigid, crescent-shaped ones.

When the red blood cells are mostly sickle shaped, sickle cell disease is the result. Its victims suffer from attacks of pain, stroke, and shorter life expectancy. However, there is one benefit. Malarial parasites do not reproduce well in sickle-shaped red blood cells, because the malaria parasite cannot hijack the host cell's supply of a protein called actin. The malaria parasites must co-opt the actin to reproduce in the host's red blood cell. Under ordinary circumstances, actin helps keep red blood cells disk shaped. In the sickle cells, the actin supply is disrupted, so the malarial parasites suffer. The trade-off is that the sickle shape of the red blood cells causes its own problems.

Sickle cell is a recessive genetic disease. People who inherit two sickle cell hemoglobin genes, one from each parent, have sickle cell disease. Those who inherit only one sickle cell hemoglobin gene from one parent and one normal hemoglobin gene from the other parent do not develop the disease. Yet, in malaria-endemic regions, those with even one copy of the sickle cell hemoglobin gene are still better at surviving malarial infections than those with two normal hemoglobin genes.

So people with one of each type of hemoglobin gene are in the Goldilocks zone—they avoid sickle cell disease *and* can fight off malaria infections. These heterozygous individuals have higher survival than do those with normal red blood cells. People with two normal hemoglobin genes are more prone to die young of malaria, and those with two sickle cell genes are more prone to die from complications of sickle cell disease. Evolution by natural selection struck a bargain between resistance to malaria on the one hand and susceptibility to sickle cell disease on the other.

Malaria is so deadly in infants and young children that the sickle cell hemoglobin gene has spread by natural selection very rapidly in equatorial Africa, parts of the Indian subcontinent, and the Arabian Peninsula, where malaria is endemic. The proportion of people carrying the sickle cell gene variant is as high as 20 percent in some of these populations.

Enter the cassava, which makes cyanogenic glucosides like those found in apple seeds. When a cassava tuber or an apple seed is wounded by herbivores, whether through the mandible of a beetle or the molar of a human, cyanogenic glucosides are converted into toxic cyanates like hydrogen cyanide in the digestive tract. Because of this toxin, people remove as much of the cyanide-producing toxins as they can from cassava by soaking the tubers in water or spreading them out in the sun. Yet, even highly processed cassava still contains some residual cyanogenic glucosides, and this is where things get interesting.

Cassava was introduced to Africa by Portuguese colonists from South America around 1600 CE and quickly became a primary carbohydrate source in the equatorial region of the continent. Fatimah Jackson found that in the northwestern and western part of Liberia, which she investigated as a case study, people who use cassava seasonally ingest, on average, nineteen milligrams of cyanates per person per day. In southeastern and central Liberia, cassava is used year-round, and the per-capita cyanate intake rises nearly fivefold, to ninety-five milligrams daily.

Two activities of cyanates in the body may influence both malaria resistance and the severity of sickle cell disease. First, at moderate levels, cyanates may kill malarial parasites or inhibit their growth in the blood. Second, even at relatively low levels, the chemicals can bind to the sickle cell hemoglobin protein and, in so doing, actually change the shape of those cells back to disks—in other words, the cyanates can "de-sickle," at least in the laboratory. Scientists have even examined whether cyanates at the right level might be a way to treat sickle cell disease. In light of these two observations, Jackson argues that cyanates incidentally consumed in the diet may confer some direct malaria protection on the one hand and alleviate symptoms of sickle cell disease on the other.

Putting two and two together, Jackson proposed that in the roughly four hundred years since the introduction of cassava in Liberia from South America, the reliance on this starchy plant and the incidental intake of cyanates threw a wrench in the war between humans and malaria. Such a time frame—only eighteen human generations since the arrival of cassava into what is now Liberia—is an even more recent time frame than the rise of the sickle cell gene in response to evolutionary pressure from malaria.

In Jackson's view, the levels of cyanates in the diet and the frequency of the sickle cell disease in the population could reflect a cost-benefit analysis that depends on malaria prevalence. Under her theory, we expect higher cassava consumption rates to lead to higher cyanate levels in the blood. Higher blood cyanate levels would in turn drive down both malaria prevalence and sickle cell disease. The patterns Jackson observed across Liberia were largely consistent with this prediction.

For Jackson's idea to explain the patterns she found, the cyanates would have to make it into the bloodstreams of infants, who are the most vulnerable to malaria because they are the most likely to die from malaria. Indeed, cyanates can cross the placenta and are found in human milk. Infants are even weaned on cassava in Liberia, and newborns there are fed cassava as a clan initiation rite. Cyanates, it seems, may serve as unintended prophylactics for the most vulnerable to malaria in Liberia.

However, cyanates didn't evolve for our benefit. Proof of this comes from konzo, or spastic paraparesis, a disease caused by chronic ingestion of high levels of cyanate from cassava. Konzo causes irreversible paralysis and occurs across equatorial African communities that eat cassava. Drought increases cyanogenic glucoside levels in cassava as tubers persist in the ground, and certain harvesting practices, like pruning, can also inadvertently raise levels. During war, as journalist Amy Maxmen has reported, insufficiently processed cassava and the preferential harvesting of sweet varieties by invading militias leads to increased cyanate levels in the plant and a concomitant increase in the incidence of konzo.

Although the benefits of cyanate consumption when malaria is present may not fully outweigh the costs—the risk of developing konzo—Jackson's hypothesis may be recapitulated in the native range of cassava in South America. The Tucano people of northwest Amazonia rely on the high-cyanate cultivar of cassava (bitter cassava) even though a low-cyanate cultivar (sweet cassava) is available to them. Why the Tucano people prefer the more cyanogenic variety is a mystery. Yes, the bitter cassava may be more resistant to pests, but the two varieties show no differences in yield, so agricultural productivity doesn't explain the Tucano preference. The real reason may have to do with malaria too, although this idea is just speculative.

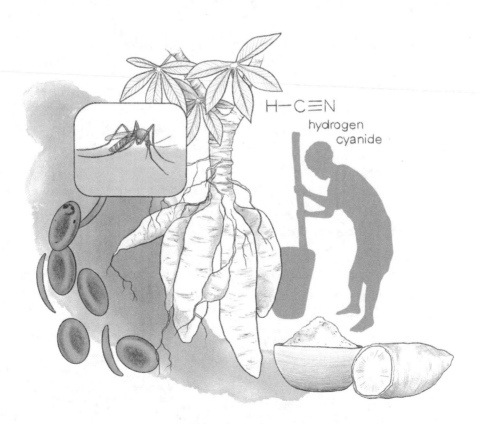

H—C≡N
hydrogen
cyanide

Malaria was introduced to the Americas from Europeans and enslaved Africans about a hundred years before cassava was moved in the other direction, from Amazonia to equatorial Africa, in 1558. As a result, malaria became endemic in the Indigenous communities of the Amazon. Indigenous Amazonians, however, do not carry the sickle cell hemoglobin gene variant that protects from malaria. In evolutionary terms, this variant was not helpful to its carriers, since these human populations evolved in isolation from the disease for thousands of years.

Under this idea—which is just that, an idea—once malaria was introduced into Amazonia at the time of European contact, high cyanate levels ingested from bitter cassava may have provided the Tucano people some protection from malaria. The lesson is that the Tucano people might have known what they were doing with bitter cassava. Although this possibility is a hypothesis, we have learned that human cultures many times independently evolve to use the same natural toxins present in the organisms around them.

Jackson's idea is that the three-way interaction between our genes, the incidental ingestion of dietary toxins, and high mortality caused by a deadly parasite changed the trajectory of human evolution. Like so many other public health concerns affecting the Global South, Jackson's compelling hypothesis has not been given the attention it deserves by the scientific community.

A parallel interaction, this one between a different dietary toxin, our genes, and malaria risk gives additional credence to Jackson's hypothesis about how relationships between food, toxins, and infectious disease can affect human evolution. Just before the COVID-19 pandemic unfurled, Shane and I flew to Rome. We rented a room at the stunning sixteenth-century Villa Medici, now the French Academy in Rome, for a few nights before I attended a conference. Let's just say that it was a room with a view, and it wasn't expensive to rent! Although we didn't want to leave the villa on our Roman holiday, we had to eat. Only a few blocks away was a wine bar where we stopped for a glass of prosecco and some antipasto. Just as we sat down, a plate of appetizers, including some fava beans, arrived at our table.

Inspired by how much we enjoyed the buttery and nutty flavor of the fava beans, when we returned home to Oakland, we planted some in our garden beds as a winter cover crop a few months before the summer tomato-growing season started. Fava beans, like soy and many other beans, are nitrogen-fixing: the symbiotic soil bacteria they house in their root nodules release homegrown ammonia fertilizer into the soil.

The bacteria living in the root nodules of many legume plants, including fava beans, require an environment with low oxygen to do their magic of converting nitrogen gas from the air into ammonia. To reduce oxygen levels in the root nodule and help the bacteria along, the plants deploy their own blood-red hemoglobin. The plant-made hemoglobin binds to oxygen and eventually lowers the oxygen levels in the root nodule. Think of this nodule as a tiny chamber wherein the bacteria can fix nitrogen. The plant creates this safe environment for the bacteria, and in return, the plant gets to use the ammonia the bacteria make as fertilizer. Incidentally, a similar hemoglobin from soybeans gives some vegetable-based "meat" both its meaty taste and its red color. The meaty taste of beef is partly due to the taste of hemoglobin.

The fava beans we later harvested were delicious when sautéed in a little olive oil. I can see why they became a staple in so many different parts of the world. Luckily, neither Shane nor I has the hereditary condition known as favism—the most common inborn enzymatic deficiency in humans. If we did have it and had had a severe case, we would have found out in Rome, the first time either of us had eaten fava beans, and we probably would not have planted the beans in our yard.

Favism is caused by one or more naturally occurring mutations in a gene on the X chromosome. This gene encodes the human glucose-6-phosphate dehydrogenase (G6PD) enzyme, which is needed to produce the antioxidant nicotinamide adenine dinucleotide phosphate (commonly known as NADPH). The antioxidant protects our red blood cells from the normal oxidative damage they endure as they race around our bodies delivering oxygen.

Favism is exacerbated by eating fava beans, which contain the alkaloids vicine and convicine. These alkaloid prodrugs break down in the gut into the toxins divicine and isouramil and trigger the rapid death of red blood cells in people carrying the mutated G6PD gene copies. The sudden death of red blood cells causes jaundice, an acute loss of red blood cells as the spleen removes the dead cells from circulation, anemia, and a suite of other symptoms, including high fever.

Paradoxically, the proportion of people with favism is highest in regions of sub-Saharan Africa, the Mediterranean, the Middle East, the Caucases, Central Asia, and Southeast Asia, where fava or broad beans are a staple. People of Sephardic Jewish or Sardinian ancestry have the highest incidence of favism.

Paradoxically, the incidence of favism is highest where both fava bean consumption is high and where the G6PD genes causing favism are prevalent. This apparent contradiction can be resolved when we consider that these regions had historically high levels of malaria.

Intriguingly, the mutant G6PD enzymes that cause favism also create an environment so oxidative that malarial parasites don't do well in these red blood cells. What's more, the overall number of red blood cells (potential homes for the parasites) is reduced because they are rapidly removed by the spleen. People carrying the G6PD mutations are therefore better protected from malaria, just like those who carry the hemoglobin mutations that cause sickle cell disease.

But a complication emerges when the red blood cells are exposed to oxidative toxins like the vicine alkaloids. One theory is that people with the malaria protective G6PD gene copies that cause favism may become even *more* resistant to malaria *after* they eat fava beans. The bean-gene combo could come to the rescue when malaria is present because the fava bean alkaloids that break down into free-radical-generating toxins create an even more inhospitable environment for the malarial parasites than just the G6PD mutations do on their own. When the alkaloids are present, the G6PD-deficient red blood cells cannot neutralize the free radicals, and even more red blood cells are removed by the spleen. This means that any malaria parasites in those cells are killed in the process too.

But the cost of this benefit of favism is also quite high. The downside, of course, is that babies with favism are at increased risk of illness and even death from favism.

In Book 8 of *Lives and Opinions of Eminent Philosophers*, the ancient Greek writer Diogenes Laërtius tells the story of the death of Pythagoras—yes, the Pythagoras from your geometry class—in what is now the province of Crotone in Italy. Although all members of the cult Pythagoras led were supposedly vegetarian, the fava bean was forbidden. More generally, fava beans were seen by the ancients of the Mediterranean as harbingers of death. As Laërtius tells the tale, after Pythagoras refused to support a plan for democratic governance, the locals chased him out of town. As he fled, Pythagoras came upon a fava bean field in full bloom and would go no further. So seriously did Pythagoras take the fava bean taboo that he chose to surrender and be killed rather than enter the field and come into contact with the fava beans. However apocryphal, this tale holds a kernel of truth.

Malaria is most virulent in infants and young children, and as you might have guessed, favism affects them more than it affects adults. In the middle of the twentieth century, mortality rates in infants and young children with favism varied between around 2 percent and even as high as 8 percent before the availability of blood transfusions.

Without the antimalarial benefits conferred by the G6PD mutations, natural selection would no doubt have quickly removed the mutant G6PD gene from the population. Instead, these harmful genetic variants and the ones that cause the sickle cell trait spread and have been maintained in different human populations over the past ten thousand years.

As happens with the cyanates in cassava, mothers who consume fava beans can and do pass the vicine alkaloids to their infants through the placenta and in breast milk. This transmission sometimes triggers favism in babies. Over the long term, dietary choices and the frequencies of particular gene variants in human populations may be explained, at least in part, by evolution in the face of strong natural selection from infectious diseases.

Although cyanates from cassava and vicine alkaloids from fava beans may incidentally help protect infants and young children from malaria, some of the toxins we avoid as children are without any doubt intentionally used by adults to combat our own demons, whether physical or mental. To further understand how basic biological differences, whether across age, sex, or pregnancy status, influence our toxin intake, we must compare and contrast the diverse diets of species of our own kind: the mammals.

Between the Devil and the Deep Blue Sea

One evening, when I was around five years old, I asked my dad if I could try a taste of his beer. He agreed, but only if I took a tiny sip.

The ethanol in the beer burned my mouth and throat. But I also noticed the bitter taste. The bitterness came from the phenolic humulone (alpha-lupulic acid) from hops, the flowers of the hop plant. Lips curled, brow furrowed, I asked, "Why do you like that?" He replied, "It's an acquired taste." It was anathema to me.

Hops from the hop plant, which is in the Cannabaceae (the same family to which the genus *Cannabis* belongs), are used to help preserve beer. Some of us come to tolerate or even enjoy the bitterness of hops for its own sake or because it is associated with the delayed psychoactive effect from the ethanol—just like the bitterness of coffee and tea portends the delayed bump in mood from the caffeine. The phenolic humulones in beer inhibit the contamination of beer by the few bacteria (like *Lactobacillus*) able to invade the ethanol, the "toxic larder" that yeast produce as food and fortress.

My dislike of the bitterness of beer and my dad's love of it require more of an explanation than simple associative learning. Science demands a more general theory that can explain humans' complex and contradictory relationship with toxic natural products they ingest for sustenance, medicine, transcendence, and fun. To do that, let's wade into some features of animals' toxin detection systems.

All animals, from fruit flies to humans, can perceive potentially poisonous substances in food, drink, and air. The human nervous system is replete with receptors to do so. The receptors line the nose, mouth, and digestive and respiratory tracts and are even found in the brain itself. This chemical detection system uses a set of odorant receptors in the nose and a set of pH, bitter, salt, sweet, and umami taste receptors in the mouth. The mouth is also home to equally important somatosensory receptors that can detect both the variation in temperature and the presence of chemicals like capsaicin from red pepper, menthol from mint, and isothiocyanates from mustards.

A backup system made of some of these same taste and somatosensory receptors lines our digestive tracts, and this system communicates with a specific region of the brain called the area postrema (AP). The AP is in the brain stem, an evolutionarily ancient part of our brain. When nerve receptors in the gastrointestinal tract detect toxins, they communicate with this part of the brain to stimulate vomiting.

Our brain is protected from many toxins by the blood-brain barrier, a layer of special cells that seal off the brain from the rest of the body. It keeps the bad stuff out and lets the good stuff in, up to a point. Because the blood-brain barrier around the AP is weak, the AP can trigger vomiting when it detects toxins in the blood and spinal fluid. The AP's early warning system triggers involuntary reflexes that make us stop eating or drinking. The next involuntary reaction is nausea, followed by vomiting, which empties the stomach.

We then avoid toxins that produce nausea through conditioned taste aversion—we learn to associate a particular stimulus, say, an ingredient in food or drink, with a negative physical experience. Once bitten, twice shy, as the saying goes.

While nearly all toxins impart a bitter taste, the converse is not true: most bitter-tasting chemicals are not especially toxic. Our body's main poison detectors seem awfully risk-averse: if our sensitivity to bitter chemicals is too high, we may end up avoiding nutritious, nontoxic foods like fresh fruits, mushrooms, and vegetables. If it is not sensitive enough, we might consume harmful amounts of a poison and get sick or even never live to learn the lesson. Like all animals, we balance healthy risk aversion with the need for sustenance. We're able to do this quite well, mostly ducking unintended sickness or death by avoiding toxins, although how well we avoid toxins depends on their availability in our environment, as shown by the prevalence of konzo in people with chronic exposure to cyanate from eating certain types of cassava.

Physiologist John Glendinning has compared bitter-taste thresholds and toxin tolerance among mammal species. He proposes that we're able to thread this needle because both our bitter taste thresholds and our ability to tolerate dietary toxins were exquisitely sculpted by evolution in a way that corresponds to the likelihood that we will encounter bitter or toxic food or drink. He arrived at this conclusion by comparing different mammal species that eat different things.

He found no evidence that our perception of the bitterness of a chemical serves as a good proxy for its toxicity. To the contrary, Glendinning found no correlation between our perception of the bitterness of caffeine, quinine, nicotine, and strychnine and their respective toxicities.

Nicotine, by far the most toxic of these four chemicals when ingested, is less bitter to us than caffeine, the least toxic of the four. Nonetheless, caffeine is perceived as the most bitter by a wide margin. Omnivorous rodents like mice showed the same pattern, so there's nothing uniquely human about our poor ability to link bitterness with toxicity in taste tests. Evolutionary differences in toxin tolerance among mammal species influence how sensitive a given animal—be it human or otherwise—is to toxic dietary chemicals. These differences reflect the interaction of nerve cells that start in our tongues and end up in our brains, and they are key to understanding how humans have harnessed toxic tools.

Although I've focused much of my prior discussions on herbivorous insects, many mammal species, from rabbits to elephants and pikas to wildebeests, are also herbivores and fungivores. Because we are mammals, too, understanding how different mammals cope with nature's toxins can shed light on our own biology.

Glendinning conducted a broad survey across mammalian species that he lumped into four categories: carnivores (e.g., cats); omnivores (e.g., humans); and two groups of herbivores: grazers (e.g., sheep) that eat herbaceous plants and browsers (e.g., deer) that prefer the leaves of shrubs and trees. Of the four, carnivores are exposed to the fewest dietary toxicants, followed by omnivores, then grazers. Browsers have the highest exposure to dietary toxins.

By measuring each mammal group's response to quinine in food as a generic bitter chemical, Glendinning found that carnivorous species were the most sensitive and had the lowest threshold for rejecting quinine-laced food. In other words, it took very little quinine to get the meat eaters to ignore their food after taste-testing it, but it took a lot of quinine to get the browsers to stop eating food laced with it.

Sensitivity to bitterness was lower in omnivores, lower still in grazers, and lowest overall in browsers. So, omnivores had a higher threshold for bitterness than carnivores, grazers tolerated even more bitterness, and browsers tolerated the most bitterness overall. These rankings align with the known occurrences of plant and fungal toxins in the diets of each group. Carnivores, for instance, which seldom eat plants or fungi, generally encounter far fewer dietary toxins in their daily lives than herbivores do. Perhaps, therefore, the low bitter-rejection threshold of meat eaters evolved to protect them from their poor detoxification repertoire—their instincts reflect their day-to-day diets.

Considering this hypothesis, we'd expect grazers and browsers to have adaptations that allow them to ignore the same chemicals that would trigger rejection by carnivores and omnivores. This premise is borne out by observation, and the way it works brings us back to the tannin-binding salivary proteins we discussed in chapter 2.

Tannins are generally found in the highest levels in shrubs and trees, which are targeted by browsers like deer. Browsers have more tannin-binding salivary proteins than grazers, omnivores, and carnivores have. By binding to dietary tannins as the animal eats, these salivary proteins appear to sequester the tannins away from the bitter receptors, improving the taste of the leaves while also preventing the tannins from wreaking havoc inside the body as toxins.

Habituation is another way animals that frequently encounter toxins overcome the distastefulness of bitter foods. Animals that are repeatedly exposed to the same toxin can, over time, become less sensitive to it. There is evidence that browsers can habituate more readily than grazers in this way. Finally, animals with higher bitterness thresholds are generally better equipped to neutralize plant toxins than are animals with lower bitterness rejection thresholds. And like the oak gall wasps that may recruit tannins from the oaks to protect their grubs from pathogens, sheep infected with worms eat more tannins than do uninfected sheep, and the extra tannins reduce worm burdens. So, the ability to sense the bitter compounds is still important, even for highly tolerant animals, for the purposes of self-medication.

In accordance with all this, it usually takes a smaller dose of a given toxin to incapacitate a carnivore than it does to do so for an omnivore or a herbivore. Carnivores tend to have a sparser genetic toolkit than herbivores have to detect and detoxify dietary poisons.

As any cat or dog owner knows, Glendinning's findings support the fact that dogs and cats can be poisoned by many plants and fungi that pose no real problem for omnivorous humans or herbivorous pets like rabbits. For example, grapes, raisins, and tamarinds contain high levels of tartaric acid, which can cause acute kidney failure in dogs when ingested. Another well-known example is from plants in the genus *Allium*, including chives, garlic, leeks, onions, ramps, and shallots. Such plants are highly toxic to both cats and dogs owing to disulfides, which cause anemia, partly by reducing the activity of G6PD in a process similar to what occurs in favism. These same plants, however, are not toxic to humans, hamsters, rabbits, or horses — all omnivores or herbivores.

You may be wondering how these comparisons across animals and their relationships with nature's toxins apply to us. When we instead compare bitter thresholds and detoxification repertoires across the developmental stages of our own species, a remarkably parallel set of patterns emerges with those Glendinning found across mammal species with different diets. Before we do that, we need to briefly retrace our own evolutionary history.

Sugar and Spice

Few children like to eat their broccoli. For me it was not only broccoli but also carrots and any other fresh vegetable, including celery, lettuce, onions, and tomatoes. Spicy foods fell into the same category: avoid. If my brother and I were eating at the kitchen counter, my goal was to wedge the carrot sticks under a cable stapled under the counter. Once dehydrated, they would fall to the floor and be eaten by the dog.

Omnivorous dogs like to eat fruits and vegetables, but meat-eating wolves, the ancestors of dogs, typically don't. Wolves in Minnesota, however, are known to seasonally feast on blueberries at times of the year when animal prey are scarce.

As they began to associate with us, dogs rapidly evolved to make far more of the starch-digesting salivary amylase enzymes in their mouths than their wolf ancestors did. The same increase happens in human populations that rely heavily on starchy foods but not in groups that don't—another case of convergent evolution. We have adapted to the predominant foods in our societies—and dogs adapted to these same foods, too.

Despite our shared ability to predigest starches, dogs are far more sensitive to the harmful effects of dietary toxins than we are. The American Society for the Prevention of Cruelty to Animals lists over four hundred species or varieties of plants that are toxic to dogs. This difference between dogs and humans may exist partly because managing ingested toxins is a far more complex process than simply predigesting starch in the mouth. Dogs began to diverge evolutionarily from carnivorous wolves only around forty thousand years ago and may have been domesticated by around twenty thousand years ago. Unlike us, they did not evolve from herbivorous ancestors tens of millions of years ago. Very little evolutionary time has passed for dogs to evolve novel detoxification systems. As with our Achilles' heel, evolution uses what is available, not the ideal, as populations adapt to a changing environment.

In contrast, our own toxin coping mechanisms are superior to our dogs' mechanisms in part because our lineage evolved from ancestors already good at neutralizing plant and fungal toxins. Although our first primate ancestors arose from insectivorous shrewlike animals that lived in trees about fifty-five million years ago, the anthropoid lineage, our own, arose about twenty-five million years ago, and these ancestors were probably herbivorous. Scientists think that our most recent common ancestor with chimpanzees (an ancestor that lived some five to ten million years ago) was also herbivorous and was, specifically, a tree-dwelling fruit specialist.

Our bodies evolved from primate ancestors who were herbivorous for tens of millions of years. Thanks to them, at least in part, we have better systems for detoxifying dietary toxins than do dogs, which evolved from their carnivorous ancestors far more recently.

Then something remarkable happened around 4.4 million years ago—as discovered by anthropologists Tim White, Gen Suwa, and Berhane Asfaw. An ancestor of all hominids, *Ardipithecus ramidus*, evolved the ability to both swing from trees and walk upright on the ground using two legs. Our first truly bipedal ancestors, the australopithecines, then permanently dropped from the trees to the ground in Africa two million years ago. These creatures evolved into the first humans, *Homo erectus*.

Although the *H. erectus* brain was not as big as ours, this species had larger brains, smaller teeth, and less powerful jaws than their immediate ancestors had. These differences point to a shift in diet.

The brain size of *H. erectus* increased rapidly until around a hundred thousand years ago, when it reached the range of our brain sizes. When we compare the modern human brain to that of our closest living relatives, the chimpanzees, ours is three times larger overall, our cerebral cortex two times larger, and the need for glucose in the brain twice as great. Constituting only 2 percent of our body weight, our brains use 20 percent of the glucose we consume. The culprit is the sugar-hungry cerebral cortex, the wrinkly outer layer of our brains, the part that really makes us human. Abstract thought, language, long-term memory, and the sense of self are all seated in the cerebral cortex.

There is some evidence that as our brains got bigger with evolution, the size of our guts seemed to shrink in tandem. One possible explanation for this is the *expensive tissue hypothesis*. It holds that the evolving human body traded disinvestment in digestive tissue for investment in brain tissue. In support of this hypothesis, primatologist Richard Wrangham and his collaborators proposed that our unique ability to cook using fire—essentially outsourcing some of the work of digestion to the cooking process—allowed the human brain to reach its current size. Wrangham's "we are because we cook" theory proposes that cooking unlocked access to stores of carbohydrates needed to support what were the largest cerebral cortexes of any animal species: our own. What was for dinner (and breakfast and lunch)? Quite possibly, the abundant underground starchy tubers that, like yams, are ubiquitous in the cradle of human evolution, Africa.

As the brains of *H. erectus* evolved to be larger and larger over time, the species also evolved smaller differences in body size across sexes. At the same time, larger, more complex social structures evolved; division of labor and overlapping generations arose; and life spans were extended. In other words, these ancestors of ours became increasingly similar to us. Wrangham and collaborators argue that the switch from eating raw to cooked plant foods, and especially starchy tubers, was the catalyst for this transformation.

More important for the purposes of this book, our (possibly) shrunken guts may have also meant that early humans could no longer persist on the high-fiber, raw plant diet of our ancestors. To extract energy from all that fiber, long gut passage times and large gut surface areas are optimal. Also consistent with Wrangham's hypothesis is that cooking is one of a host of techniques we still use to deactivate dietary toxins. Other ways of removing toxins include fermentation, adsorption using clay, drying, soaking, physical processing, and using ashes or other caustic material to change the acidity of a substance.

The extraordinary energy needs of the human brain also help explain why children are so fond of fruits and sweets. Even though they are smaller than adults, glucose uptake in the brains of children is twice as high as in adults. Kids like sweet things for other reasons, too, but one reason is that their brains need a large supply of energy. Once the association is made between that need and sugar, their dopamine-based mesolimbic reward systems help drive them to seek it out, sometimes incessantly.

However, not all people are alike in their desire for sweets. Some children and adults really like them, and others don't. There are cultural, learned, and heritable factors involved in these differences.

Although people show much variation in their desire for sweets, each of us can be placed into one of two broad categories: sweet-liking and sweet-disliking. To tease out these differences, food scientists test people's responses to a highly sweetened sucrose drink. The sweet-liking people love it, and the sweet-disliking people loathe it. Which bin you fall into appears to be largely innate—it can be discerned in infants shortly after birth and is highly heritable. As we might expect, there is a possible connection between having a sweet tooth, the risk of developing AUD, and the degree of craving in nondrinkers with AUD.

One study found that those with familial history of AUD are 2.5 times more likely to be in the sweet-liking category than those without a history of AUD in the family. Another found that sweet-liking people with AUD took ten times as long to abstain from drinking ethanol than did sweet-disliking people with AUD. Many other data sets besides these show a link between the risk of AUD and having a sweet tooth, but some studies did not find a link.

An underactive endogenous opioid system in the brains of those with a sweet tooth may help explain a possible connection with AUD. Because ethanol releases endorphins, which activate the opioid receptors, sweet-liking people may be more prone to having an endorphin or opioid receptor deficiency and, if so, are therefore highly motivated to stimulate that pathway.

In support of this idea, sweet-likers with AUD were more responsive to the use of the opioid receptor blocker naltrexone in a small clinical study— they craved ethanol less when given the drug than did sweet-disliking people with AUD. This finding suggests that when you take the opioid pathway out of the equation with a temporary "surgical strike" using naltrexone, people more easily abstain from drinking, but only if they are sweet-liking. These results point to a connection between innately lower endorphin or opioid receptor levels, the propensity for liking sweets, and the susceptibility to AUD or other drug use disorders. Note, however, that these studies are small and preliminary.

People I have known who struggled with alcohol dependency had a sweet tooth, and when abstaining, they managed cravings using sugared drinks and foods. You probably also know people just like this. However, we should put the results of the previously mentioned studies in perspective. Just because somebody likes sweets doesn't mean they are or will become dependent on a drug of reward, and many people with drug use disorders dislike sweets. These are correlational studies, and drug use disorders have highly complex causes.

Still, some people no doubt use natural toxins to modulate their brain activities to feel more "normal." In the next chapter, we will explore what factors besides having a sweet tooth might cause some people to slip off the knife's edge and become dependent on these drugs. But for now, the next sections examine why bitter and spicy food tends to be disliked at certain times in our lives and liked at other times.

Growing Out of It

In search of glucose, I eagerly ate *cooked* vegetables as a child, especially carrots and potatoes. They were starchy or even sweet—and definitely not bitter. Like most other people, I grew out of my aversion to fresh vegetables.

To understand why kids might like sweets but not bitters and why most adults overcome childhood aversion to bitters, we need to examine the differences between individuals in various stages of life. In doing so, we will see how the risk of intoxication seems to govern differences in dietary toxin aversion and consumption between children and adults, sexes, and those pregnant and not pregnant.

Anthropologists Edward Hagen, Roger Sullivan, and their collaborators have used these differences to help develop the *neurotoxin regulation hypothesis.* This theory has changed the way I look at our relationship with nature's toxins. It partly hinges on the World Health Organization's World Mental Health Surveys on the human use of four of the most common psychoactive drugs derived from natural toxins: alcohol, tobacco, cannabis, and cocaine.

A total of 54,069 people from seventeen countries participated in interviews using the same questions. All were asked if they had ever used any of these four psychoactive drugs and, for nine of the countries, when they first used. Among the findings: Male participants were more likely to have used than females were, and while 74 percent of US participants reported that they had used tobacco, only 17 percent of Nigerians did. Around 4 percent of Colombians reported trying cocaine, whereas 16 percent of those in the US survey did.

One striking pattern that held across all seventeen countries emerged: the relationship between drug use and age of onset for the four drugs of abuse. The age of onset refers to the age in years when participants first used the drug, if at all. The median age of onset was 16 to 20 years for alcohol, 16 to 21 for tobacco, 18 to 22 for cannabis, and 21 to 24 for cocaine. In other words, most people who used these substances did so first in the middle to late teenage years for most legal drugs and in the early to mid-twenties for illicit drugs.

Perhaps the most salient finding is that children generally do not use any of these psychoactive drugs. There is virtually no reported use of any of the four before the age of ten. Hagen and his team propose that a "developmental switch" gets flipped as we make the transition through adolescence and begin to experiment with psychoactive drugs of reward.

The lack of childhood drug use isn't just based on survey data, which can be biased in many ways. Fortunately, blood levels of cotinine, a metabolite of nicotine, don't suffer from the same pitfalls that survey data faces. Cotinine levels neither hide the truth, misremember, nor forget. The data from a study of 18,382 children, adolescents, and young adults from 1999 to 2010 in the United States tell the same story. Although some of the children tested positive for cotinine, the levels of cotinine in their bodies were all within the range produced by secondhand smoke. There is simply no good evidence that children below the age of ten use tobacco. But by age eleven, the number of children with cotinine levels consistent with tobacco users and smokers in particular began to increase dramatically.

I know what you are probably thinking—younger children don't have the same kind of access to tobacco that younger adults do. Yet there is virtually no coffee consumption in children up to the age of fourteen, either, and coffee is far more widely available.

The best explanation for this pattern is that, like carnivorous animals, children up to the age of ten are far more sensitive to bitter chemicals and are thus more likely to reject bitter-tasting food or drink than adults are. Children have higher densities of taste buds than adults have, and up to adolescence, youngsters reject new foods at a much higher rate than adults do. A child's reluctance to try a new food is part of a sensibility called neophobia, or the fear of things new. As babies reach their first year and are often being weaned off milk, they are beginning to eat foods that may contain natural toxins. At this point in their lives, their ability to detoxify particular drugs ramps up and can exceed that of adults. This development is consistent with an explanation that intertwine cultural and biological factors: the chemical detection systems and detoxification systems of growing children have been tweaked by evolution to protect the rapidly dividing cells to a degree that reflects the risk of exposure to dietary toxins.

Intriguingly, researchers have also found that premenopausal women are more averse to bitter-tasting food and drink and are better able to detoxify most drugs than men of the same age are. Overall, women have more taste buds than men do, and the expression of detoxification genes is higher than in men. Again, we see a similar pattern in young children. This pattern reflects the risk to the developing body, which spends its first nine months inside the body of another.

The vast majority of pregnant people surveyed reported disliking certain foods and drinks and experiencing nausea and vomiting during pregnancy. Many of you who have been or are pregnant have probably had your own experiences with quirky, surprising dietary dislikes during pregnancy. Some aversions include caffeine, alcohol, and tobacco. Bitter dietary chemicals are the least tolerated during the first trimester, and the body's ability to detoxify drugs generally also increases during pregnancy.

For example, as discussed earlier, low doses of caffeine appear safe for most people. But high doses consumed in the first and second trimester of pregnancy can increase the risk of miscarriage, low birth weight, and pre-term birth. Although caffeine may not be a strict teratogen—a chemical that causes developmental abnormalities in fetuses—it can still pose big problems in high doses during pregnancy.

Thalidomide was an infamous teratogen prescribed in the 1950s and 1960s to—ironically—combat morning sickness in pregnant women. This anti-nausea and anti-vomiting drug caused virtually no side effects in pregnant mothers and was seen as a wonder drug for morning sickness. Unfortunately, when used in early pregnancy, unbeknownst to well-intentioned physicians and patients, thalidomide resulted in major developmental abnormalities, including severe limb malformation, in the babies born. This sad situation demonstrates why toxins that may seem benign to us as adults can be toxic to developing human bodies, whether the toxins are natural or synthetic.

Ethanol, nicotine, and many other natural toxins are teratogens. In severe cases, fetal alcohol syndrome results in physical abnormalities in the face and brain. Likewise, smoking and, sometimes, even smokeless tobacco use increase the risk of preterm delivery, stillbirth, perinatal death, and sudden infant death syndrome, as well as motor, cognitive, and sensory deficits in infants and toddlers. This is just the tip of the iceberg; many other toxins that impart innocuous effects on adults are problematic in developing humans. During development, whether before or after birth, cells are dividing rapidly. It is during this phase, from conception through puberty, that mutagenic or teratogenic toxins wreak the most havoc.

We can infer from these observations that during childhood and pregnancy, the costs of consuming psychoactive drugs outweigh the benefits. The benefit-to-cost ratio begins to flip for some drugs once our bodies are completely developed, sometime in our middle to late teenage years or early twenties.

Hagen and collaborators propose that these differences in bitter food aversion, detoxification efficiencies, and behaviors across human development, sex, and pregnancy status reflect the fact that dietary toxins exact their highest toll on the rapidly dividing cells of developing fetuses and young children. In other words, because of the increased danger some dietary toxins pose, children and people who could become pregnant or are pregnant could benefit from being more sensitive and averse to these toxins and from being able to rid the body of them more efficiently.

Now that we have at least one hypothesis explaining why we grow out of our aversion to many natural toxins that become commonly used in adulthood, there is the other side of the equation: why we seek them out as adults and become dependent on them, sometimes for the better, as with caffeine, and at other times for the worse, as with nicotine.

Growing into It

Since time immemorial, humans have carefully regulated their natural toxin intake to cope with the vicissitudes of life. The advent and wide availability of pure drugs in the twentieth century have upset this fine balance, as imperfect as it already was.

The *paradox of drug reward*, also coined by Sullivan and colleagues, describes the phenomenon wherein adults can use and even become addicted to the same chemicals that, like caffeine, cocaine, ethanol, and nicotine, evolved to serve as toxins by plants and other sessile organisms to keep enemies at bay.

At first blush, resolving this paradox might seem like a mere academic exercise, bringing little comfort to those who are addicts themselves or have lost loved ones to addiction. But everything covered in this book so far bears directly on this issue of addiction. Unraveling the revulsion-addiction paradox will not only help us better understand why people end up with drug abuse disorders but will also give us a new way to put the last five hundred years of human history in clearer perspective.

We don't become addicted to aspirin, but we do to nicotine. Both drugs are used by hundreds of millions of people all over the world every day (see the appendix online). Depending on the dose and on timing, duration, and method of dosing, both drugs can harm or even kill us. Nicotine is addictive primarily because it triggers the accumulation of dopamine in our brain's reward center. Aspirin does not. Although this difference obviously explains why we become addicted to nicotine but not aspirin, the more salient question is what motivated us to use drugs of reward like nicotine in the first place. It is clear we started using aspirin—from the Neanderthal "Sid" to Reverend Stone—to deal with fever, pain, malaise, and discomfort. In other words, aspirin helps solve a very practical problem. Does this purpose also hold for drugs that we now see largely as recreational, like nicotine?

Humans have been using nicotine for thousands of years, along with arecoline (described below), caffeine, cathinone, cocaine, ethanol, ephedrine, morphine, and THC. Most of these drugs mimic the structure of a neurotransmitter or enhance their levels in the brain through other means.

Of this list, the alkaloid arecoline is new to our discussion, but it is one of the most ancient in terms of human use. What's more, it remains number four on the list of the most widely used addictive drug in the world, only behind nicotine, ethanol, and caffeine (see the appendix online). Arecoline is found in betel nut, the seeds of the *Areca catechu* palm grown in South

Asia, many smaller islands in the Indian Ocean, Southeast Asia, and Oceania. Betel nut is often mixed with lime or ash, much as coca leaves are in the Andes and parts of Amazonia, and rolled into the leaves of a spicy plant from the black pepper family called *Piper betle*. The mixture is then chewed as a quid like both coca and pituri quids. A vasodilator and stimulant, arecoline produces a euphoria similar to that caused by nicotine, although arecoline binds to muscarinic acetylcholine receptors, not nicotinic acetylcholine receptors, so the body's responses are physiologically distinct.

The oldest bona fide evidence of betel chewing comes from a burial pit in Duyong Cave on the island of Palawan in the Philippines. The teeth of the individuals were stained red by the betel nut they chewed more than six thousand years ago, just like the scarlet teeth of betel chewers today. Amazingly, six clamshells were also found with the skeletons. These were used as vessels to hold lime, as they still are today.

Betel nut, black and green teas, cacao, cannabis, coffee, coca, ephedra, ethanol, guarana, guayusa tea, khat, opium, tobacco, yaupon tea, and yerba maté were all historically viewed as much as food or drink as they are daily medicine by those of us who use them. All of the new terms in that list, guarana, guayusa tea, and yaupon tea, are sources of caffeine (see the appendix online).

Yet while all the drugs in these substances are addictive, for many of us who have taken our first deep drag, big sip, or generous dip, the initial experience can be simply repugnant, both in taste and in the effect on our bodies. As my dad said to me, it is an acquired taste. Much is revealed by this expression because it acknowledges how distasteful or even painful the experience can be at first.

Yet we sometimes go back for more if the brain's reward circuits are triggered; such triggering can overcome the initially repugnant experiences. The drugs change our state of mind in ways that may only happen without them in rare moments of elation, excitement, or fear. For example, many of us learn to associate the bitter taste of coffee with the caffeine reward and even come to enjoy the unadulterated taste. The same applies to these other substances. We step over that bitter or burning or nauseating bar because we know that if we do, we will end up feeling a sensation we desire, a sensation we grow to need.

Remarkably, this process evolved so many times independently, in culture after culture. It takes time to associate an initially repugnant experience with a rewarding feeling, and it also takes culturally transmitted knowledge of why and how. A source of the toxin is also required, whether it comes from a vendor, a dealer, a friend, a parent, a shaman, or nature itself.

The neurotoxin regulation hypothesis proposes that addictive substances like most of the natural toxins we take were first used more to ease our suffering than to have fun, relax, commune with the gods, or take a trip. If our ancestors first harnessed drugs of reward as medicines in the broadest sense, this practice could account for our strange but universally human behavior of seeking out psychoactive drugs on a daily basis.

As "Sid" and the bitter-pith-chewing chimpanzees taught us, self-medication is nothing new. Birds do it, bees do it, and we do it. But something quite different, something uniquely human, about our relationship with these chemicals sets us far apart from all other animals. We need to reconcile the fact that most of us take psychoactive drugs regularly and that some of us use them as a way to commune with the spiritual realm.

Under the neurotoxin regulation hypothesis, what's good for the goose is good for the gander. It argues that both aspirin-like chemicals and nicotine were leveraged by our ancestors as medicines first, but only nicotine turned out to be a psychoactive drug. Intriguingly, unlike most natural toxins we use as medicines, psychoactive drugs like nicotine make us acutely aware that we've taken them. We even get a sense of how much we've taken because they cross the blood-brain barrier, creating distinct, dose-dependent physical and psychological effects. This is the reason you can feel a second cup of coffee much more readily than you would a second tablet of aspirin.

The neurotoxin regulation hypothesis—which is still very much that, a hypothesis—can trace its roots to a remarkable idea on the origin of psychoactive drug use in humans. Ethnobotanists Eloy Rodriguez and Jan

Clymer Cavin studied the practices of diverse communities of Indigenous Amazonians, who use psychedelic plants for healing. In Iquitos, Peru, *vegitalista* shamans use psychedelic plants as *vegetales que enseñan* or "plant teachers," which impart curative knowledge to the shamans who often take the drugs along with the patient. A similar practice exists today with mescaline use in the Andes and Mexico and has been in place for thousands of years.

Rodriguez and Cavin's first point was related to a finding made by ethnobotanist Richard Evans Schultes. The psychedelic plants chosen by shamans across the Amazon typically shared the same chemical backbones, either indole (like DMT) or isoquinoline (like quinine), both of which produce psychoactive and purging (vomiting or diarrhea, or both) responses. One concoction widely used to cleanse the body is ayahuasca. Purification and purging are essential parts of the practice of using psychedelic plants.

Rodriguez and Cavin noted that the shamans did not choose these plants by chance. The blurred vision, euphoria, and hallucinations, among other neurological side effects, are intertwined with the purgative effects. All these effects help explain why the plants are used as "plant teachers" thought to help the shaman diagnose the illness, to help the patient cleanse the body through purging, and to perhaps direct the toxic effects to any parasites within, as actual cures.

At least some of the chemicals used are also effective antiparasitic drugs. Emetine, which, as previously described, contributed to the death of Karen Carpenter, is a potent anti-amoeba drug still used to treat dysentery in extreme cases, and quinine is an antimalarial drug from the bark of cinchona trees. At least in the laboratory, the ayahuasca's harmine alkaloids, which inhibit the activity of monoamine oxidase enzymes, are toxic to the parasite that causes Chagas disease.

Putting all of this together, Rodriguez and Cavin concluded that diverse groups of Amazonian peoples, spread across thousands of miles, from the Andes to the Atlantic, have repeatedly sought out plants that make psychedelic indole and isoquinoline alkaloids for medicinal purposes. Although these groups looked at the neurological effects as a way to measure the dose, they also used these toxins for both cultural and spiritual purposes. The plants, to these people of the Amazon, are both teachers and medicines.

Some of these natural toxins used by *vegitalista* shamans are neurotransmitter mimics, acting on the neurons in tapeworms and roundworms too. As a result, these toxins can serve as anti-worm drugs. Our shared evolutionary heritage with parasitic worms, as unpleasant as that truth might be, means that they are vulnerable to some of the same psychoactive drugs that we are.

Drugs like arecoline and nicotine have been used to treat domesticated animals for parasitic worms. Betel nut is a traditional Chinese medicine for tapeworm infections, the active ingredient being arecoline. Researchers in one study looked at tobacco and cannabis use by the Aka people of Central Africa to predict how heavily an individual was infected with intestinal worms. Both tobacco and cannabis were relatively new as cultivated plants in equatorial Africa, so the research was a case study of how the neurotoxin regulation hypothesis might work. A high concentration of nicotine or a THC derivative in the blood was associated with reduced worm burdens in the study. Higher levels of these drugs also reduced the odds of the person's getting reinfected the following year. Although preliminary, the findings are intriguing.

As with the psychedelics and the MAOIs used by the *vegitalistas* of Iquitos, over the long haul, the use of arecoline; nicotine; and THC, cannabidiol (CBD), and other chemicals in cannabis as antiparasitic drugs could have increased the survival rates of users in parasite-rich environments. This unintended benefit could have promoted psychoactive drug use, as cassava and fava bean consumption in malaria-endemic areas might have done for the non-psychoactive cyanates and vicine alkaloids. The ultimate result would have been the use of these drugs in the cultural mainstream.

The hijacking of the mesolimbic reward system by drugs of reward is also an essential part of why we use and abuse some drugs, but it is only part of the story. The overlooked piece is that it may have been a matter of life and death!

Finally, as discussed, many chemicals we use as drugs of reward from plants and other organisms mimic neurotransmitters or change the levels of neurotransmitters in the brain (see the appendix online). This chemical behavior has led to the hypothesis that when nutrition is poor, these dietary chemicals may augment the supply of our own neurotransmitters directly or indirectly. Many drugs of reward, like caffeine, cocaine, ethanol, ephedrine, cathinone, opioids, and nicotine, are used to cope with boredom, hunger, fatigue, and stress and to improve our brain's performance in different ways. However, the devil is in the details, and it is difficult to conceive of how this could work physiologically given the precise timing required for the release of endogenous neurotransmitters into the synaptic cleft followed by their reuptake and inactivation in neurons.

These hypotheses about the ultimate evolutionary reasons that humans everywhere tap into daily or periodic use of mind-altering chemicals from nature are still controversial. Yet, given that billions of people every day use some of these chemicals, it is easy to see the benefits these drugs can impart. Altogether, our deliberate uses of toxic chemicals as both medication and psychological reward support the neurotoxin hypothesis.

Psychoactive and dietary toxins are one thing. The next question we'll consider is whether the far more widely used chemicals found in spices might confer benefits beyond enhancing the taste of our food and drink. If so, these benefits might help explain our obsession with them and how spices changed the world.

11.

The Spice of Life

It is quite surprising that the use of pepper has come so much into fashion...pepper has nothing in it that can plead as recommendation to either fruit or berry, its only desirable quality being a certain pungency; and yet it is for this that we import it all the way from India!
— Gaius Plinius Secundus (Pliny the Elder), *Natural History*, Book 12, Ch. 15, ca. 77–79 CE

Feel the Burn

My maternal grandmother, Doris, often prepared roast beef for Sunday dinner. When she did, she also placed a jar of horseradish on the table for my father. I watched in disbelief as he ate the spicy slices, his eyes watering profusely, his skin reddening, his brow sweating. I could not fathom why he would invite such pain during an enjoyable meal.

But Doris understood, and she offered to add fuel to the fire, wandering around the table with her towering pepper mill, wielding it like a boss. With a few cranks, her flakes of black gold fell onto the plate of anybody who wanted in.

Now I find myself doing the same thing I found so inexplicable then. I am not alone. You may also seek out a bit of the nose-stinging, eye-watering, tongue-tingling, skin-flushing isothiocyanates in mustard condiments like wasabi and the sharp, throat-clenching piperidine alkaloids in cracked black pepper.

Plants in the mustard family Brassicaceae are far more than just a spicy condiment. They are staple crops, including arugula, broccoli, brussels sprouts, cabbages, canola or rapeseed, cauliflower, collard greens, daikon, gai lan, horseradish, kale, kohlrabi, mizuna, mustard seeds, peppergrass, radishes, rapini, romanesco, rutabaga, turnips, wasabi, watercress, and winter cress. Close relatives in the same lineage that produce mustard oils include capers, meadowfoam, moringa, papaya, and nasturtium. Toxic mustard oil defenses could be one reason why these plants are so successful: worldwide, there are over five thousand species that have the ability to make mustard oils.

The function of mustard oils for plants was revealed in an experiment conducted by USDA scientists in the early 1960s. The researchers ground up fresh turnip roots that they placed on filter paper at the bottoms of several small jars. They then lowered small cages of fruit flies into the jars to expose them to the vapor. Over 90 percent of the flies died within three hours. The toxic agent was 2-phenethyl isothiocyanate, a mustard oil.

The dead fruit flies in those miniaturized "shark cages" were killed not by the jaws of a super predator but by invisible mustard oil vapor emitted by the wounded turnip. Mustard oils serve as defensive weapons in the chemical war of nature. Because these toxins are also poisonous to the plants that make them, they are formed only when the plant has been wounded. The way this works is that protoxins called glucosinolates are stored in the plant's cells like bombs with unlit fuses in a bunker. Mustard plants also produce glucosidase enzymes, which are stored in different cells. These enzymes are like a box of matches. When glucosinolate and glucosidase come into contact as an herbivore begins to munch away, the fuse is lit and the bomb explodes, transforming the glucosinolate into the toxic mustard oil. This is why it takes a bit of chewing before your fresh arugula, daikon, or watercress begins to taste spicy, hot, or, as some describe it, peppery.

Most commercially available wasabi preparations are actually just horseradish with algae added so that the concoction resembles the real, and far more expensive, green wasabi rhizome. In both preparations, as well as in mustard the condiment, the fuse of the bomb has already been lit by the grinding or the grating: the spiciness of condiments like yellow mustard, wasabi, or horseradish hits us the instant it reaches our tongues. Kimchi and sauerkraut are in the same boat because the *Lactobacillus* and other bacteria that ferment cabbage can break down some of the glucosinolates into mustard oils using their own glucosidase enzymes.

When mustard vegetables are cooked, however, the heat disables the plant's glucosidase enzymes. For this reason, cooking is sort of like pouring water on the matches. The glucosinolates then move into our digestive tracts without forming mustard oils — not yet, anyway. Our gut bacteria have plans for the glucosinolates. After the gut bacteria turn the glucosinolates into mustard oils, the toxins are then broken down further into an amine and hydrogen sulfide gas. The bacteria use the amines as nutrients, but the sulfur-smelling gas bubbles are waste that find their way out of the other end of our digestive tract. Think of your microbiome and your gut itself as your personal kimchi fermenter.

Not only are mustard oils nontoxic at the levels we ingest, but they may even also promote health. At higher doses than most of us would obtain from our diet, they have potential as medicines. Mustard oils are being studied as a chemotherapy for cancer and as treatments for neurodegenerative disorders, autism spectrum disorder, and traumatic brain injuries, among other disorders.

I try to consume some mustard oils every day. They might come from the curly-leaf kale salad I routinely order from the Free Speech Movement Café at the University of California, Berkeley, Shane's seemingly infinite supply of pickled rainbow radishes from our garden, the baby broccoli I like to grill, or in the wasabi-laden soy sauce in which I dip my sushi.

My own laboratory has preliminary data showing that when modest amounts of sulforaphane, a mustard oil from broccoli, are added to fruit fly food, the flies live somewhat longer than do the flies fed plain food. Of course, I am excited about these results, but they should not be overhyped. The subjects were fruit flies, not humans, and our study was quite small.

Other studies showed similar life-span-extending effects of dietary sulforaphane, but this time, the subjects were flour beetles instead of fruit flies. The researchers also determined that the effect depended on the activation of an important protein in our cells. Called Nrf2, this protein is a master regulator of our overall detoxification response. When Nrf2 is activated by mustard oils and other similar chemicals, detoxification pathways are dialed up and toxins produced through normal cell metabolism are mopped up.

Another hint about why mustard oils may extend life span came from a study that showed how fruit flies fed rotenone survived longer if they were first fed sulforaphane. Rotenone is a potent toxin that creates damaging oxygen free radicals by interfering with proteins critical to cellular respiration. Sulforaphane may prime the natural antioxidant response of flies and thus make them better able to tolerate the oxidative assault that follows from rotenone ingestion.

Rotenone is an isoflavonoid toxin (not all flavonoids are good for you!) produced by tropical legume plants, including one of my favorite vegetable snacks, jicama. The tubers of jicama are *completely* safe to eat, but the seeds within the seed pods are incredibly poisonous, especially to insects and fish.

Indigenous people across the Global South have used rotenone-producing legumes to stun and harvest fish, from rivers of Amazonia to coral reefs in the South Pacific. When rotenone-containing plant tissues are released into the water, the toxin enters the gills and prevents the fish from using oxygen. Immobilized, they float to the surface, where they can be easily captured and safely eaten (the rotenone doesn't get into the flesh). To this day, rotenone remains widely used as an insecticide and a fish poison.

Rotenone can produce symptoms resembling Parkinson's disease in animal models and humans because of the acute oxidative stress it causes in the brain. Some studies show that consumption of sulforaphane beforehand may protect laboratory animals against the progression of these Parkinson's disease–like symptoms.

Scientists have dug deeper into the potential protective effects of mustard oils against Parkinson's disease by developing animal models of the disorder. One of these models mirrors Parkinson's disease in people who carry inborn mutations in the alpha-synuclein gene and are predisposed to develop a form of the disease called familial Parkinson's disease. In these people, the alpha-synuclein proteins lump together to form Lewy bodies in the brain. Lewy bodies also form in some spontaneous Parkinson's disease cases and in cases of Lewy body dementia, which afflicted the late Robin Williams. Lewy bodies are associated with the death of the midbrain's dopamine-producing nerve cells and may cause oxidative damage.

To better study Parkinson's disease in animals, researchers spliced the human alpha-synuclein gene into the fruit fly's genome, which normally doesn't have this gene. When they turned on the human gene in the fly's brain, the flies developed Parkinson's disease–like symptoms, including tremors, which were caused by malfunctioning dopamine-producing nerve cells in their brains, just as they are in humans with the disease. Flies are useful models because dopamine-producing cells probably evolved once in the brain of a common ancestor of humans and flies over half a billion years ago. In many ways, a brain is a brain in the most fundamental ways, whether it belongs to a fly or a flutist.

When the diet of the flies was supplemented with sulforaphane, the progression of Parkinson's symptoms was abated in the mutant fruit flies. What was more, the same thing was found in mouse models of the disease. Researchers injected a toxic form of dopamine into the same mouse brain neurons that die in the brains of people with Parkinson's disease. When these mice were fed a diet supplemented with sulforaphane, they exhibited fewer disease-like symptoms and lost fewer dopamine neurons than did control mice, just like the flies. So, a diet laced with sulforaphane prevented or slowed the symptom progression in fruit flies and mice predisposed to develop Parkinson's disease through the activation of the Nrf2 protein mentioned earlier.

Nrf2 is normally bound to a protein studded with toxin sensors in the cytoplasm. When sulforaphane enters our cells, it binds to the toxin sensors. Nrf2 then gets released and, like the Paul Revere of proteins, travels to the cell's nucleus to warn it of incoming danger. It does this because the nucleus is where DNA is held and where the production of proteins is initiated.

In the nucleus, Nrf2 binds to the transcription factors, which are bound to the DNA adjacent to detoxification genes. Once Nrf2 binds to the transcription factors, the duo is released from the DNA. Think of their release as akin to flipping on a switch. The gene next door gets switched on, inducing our cells to make more of a toxin-scavenging molecule called glutathione (GSH) and more of an enzyme called a glutathione *S*-transferase (GST). All of this starts when a toxin is sensed in the cell, causing the cell to turn on its detoxification machinery.

If GSH is like a sponge, then GST enzymes are like the hand that moves the sponge to the spill and wrings it out so that the sponge can keep mopping up the mess. When toxins bind to GSH—a key antioxidant produced in the cells of our liver—they are deactivated and then eliminated via urine.

People who die of acetaminophen or paracetamol overdoses succumb because too much of our liver's GSH becomes bound to the drug. There is then not enough GSH to deal with the toxic by-products our own bodies naturally produce just by living. The lack of available GSH is just the first in a one-two punch from acute acetaminophen poisoning. As the liver detoxifies the acetaminophen, a toxin forms and binds to other enzymes the liver needs to neutralize other toxins produced by our metabolism.

Although acetaminophen is actually much safer than aspirin at normal doses, it is the dose that makes the poison. The treatment for acetaminophen poisoning is a dose of a precursor of GSH, which stimulates the liver to produce more GSH, allowing for recovery by preventing acute liver failure.

The body's reaction to consuming sulforaphane is to ramp up GSH and GST enzyme production to rid the body of what it detects is a toxin. At levels consumed in our diet, these mechanisms for removal of sulforaphane are sufficient to prevent sulforaphane from really damaging our cells.

Yet, once the GSH and GST enzyme levels become elevated thanks to sulforaphane, the higher levels of GSH and GST enzymes scavenge *other* toxins, including those produced through normal metabolism. This indirect benefit of sulforaphane consumption, this GSH and GST enzyme boost, is the main way that sulforaphane slows progression of Parkinson's disease-like symptoms in lab animals. I know this was a lot of detail, but the triggering of our body's detoxification response by some dietary toxins like mustard oils illustrates an important concept as it relates not only to spices but also to other toxins we ingest.

Hormesis, which comes from the Greek word *hormē,* "to set in motion," is a poorly understood but widely observed phenomenon. Through this phenomenon, modest amounts of environmental stressors, like cold shock (ice bath plunge), heat shock (sauna), or spicy mustard oils, can protect us from cellular damage caused by both internal and external stressors that our bodies will encounter in the near future.

The idea is that mild environmental stressors can trigger an adaptive physiological response that soon dampens the negative effect of the stressor. The concept of hormesis is controversial because the underlying mechanisms driving it may be quite different, depending on the stressor, and their later downstream effects may or may not matter for our health and well-being. The long-term cost-to-benefit ratio of hormesis as a prophylaxis or treatment for illnesses is quite unclear and highly context dependent.

Heat-shock proteins get turned on when we sit in a sauna. Cold-shock proteins get turned on when we dunk ourselves in an ice bath. Detoxification pathways get turned on when we ingest toxins like ethanol. By turning on these stress responses, we may (or may not) be protected from other onslaughts like oxidative damage that our bodies cannot avoid because our normal metabolism produces toxins, too. At high levels, these three stressors—heat, cold, and ethanol—can also kill us either through acute toxicosis or, as happened with my father, chronically and indirectly. Dose is therefore a critical concept when it comes to understanding how hormesis might work.

Hormetic responses produce an inverted U-shaped curve, with the benefit on the vertical axis and the dose on the horizontal axis. A good example is ethanol and heart attack risk in adults, particularly men. On average, teetotalers, who consume no alcohol, or people who consume more than two drinks a day are at higher risk than those who consume one drink per day. Somehow, one drink may be more protective against heart attack than are zero drinks or more than two drinks per day. The trouble is, even one daily drink increases cancer risk, too. This one example shows how each stressor must be studied on its own terms, in depth, and in terms of overall health, not just in terms of one disease.

One terrific explanation of why this phenomenon exists at all proposes that it is an evolutionary adaptation in animals and fungi, whose diets are largely derived from living or decomposing tissues of other organisms, which include toxins. The idea is that animals and fungi evolved a surveillance system that can ramp up the production of detoxification pathways in *anticipation* of a chemical assault just over the horizon.

Cell biologists Konrad Howitz and David Sinclair coined the term *xenohormesis* (*xeno* means "stranger") to describe this adaptive response as it relates to our ability to sense that our bodies have taken in toxins produced by other organisms. In a way, xenohormesis is similar to other anticipatory responses to future stressors, whether it is the tanning response seen in light-skinned people exposed to UV light or the 6-MBOA-driven reproductive cycles of voles.

One last finding from fruit fly models of Parkinson's disease will help tie all this together. The protective effect of sulforaphane against the progression of Parkinson's disease symptoms was not found in fruit fly mutants that could not make enough GSH or GST enzymes to detoxify mustard oils. In other words, sulforaphane does not prevent flies from progressing through Parkinson's disease–like symptoms if GSH or GST enzyme levels are low. GSH and GST enzymes are therefore necessary for the protective effect of sulforaphane in flies. This matters because, as my former PhD student Andrew Gloss, other collaborators from the Max Planck Institute for Chemical Ecology, and I discovered, fruit flies and humans use the very same GSH- and GST-based pathway to detoxify dietary mustard oils. In the most fundamental of ways, we humans aren't physiologically that different from many other animals.

All these observations suggest a tantalizing idea: that dietary supplementation with chemicals found in spicy condiments—sulforaphane and other mustard oils—might mitigate Parkinson's disease progression in animals and humans. Clinical trials are required to address the question in humans, and there is no guarantee that sulforaphane will show a therapeutic effect at all for this or any other ailment.

Small clinical trials have found that sulforaphane pills reduce symptoms of autism spectrum disorder. This finding is exciting, but again, it is very preliminary. Both neurological disorders might be ameliorated through similar protective mechanisms involving elevation of GSH and GST enzymes.

Now I hope you can see why I have such a soft spot for fruit flies and, more importantly, why modern medicine owes them so much. Fruit flies helped us understand why plants make mustard oils—to protect themselves. And fifty years later, these insects were the first organisms to show why the same chemicals help the body clear away homegrown toxins in the brain. The findings in turn helped scientists prevent the progression of experimentally induced Parkinson's-like symptoms in the fruit fly. And this newfound knowledge will, hopefully, someday be applicable to humans.

The truth is that none of us think too much about the health benefits of spices like mustard oils when we use them. Yet, their health benefits may indeed be something we should consider when we reach into our pantry or fridge.

Why Spices?

The exciting possibility that a plant toxin found in spicy condiments, salads, and cooked vegetables protects against neurological disorders points to an ultimate evolutionary reason humans use spices—they enhance our health and wellness. Our food is not only nourishment but also medicine.

Before I dive further into the question of why we use spices from this perspective, I'll first address why we spend so much energy removing toxins from plants and fungi *before* we consume them. The processing of plants and fungi before consumption, whether it's done through peeling, soaking, pickling, fermenting, or especially cooking, defines what it means to be human. Why don't humans just breed plants with reduced levels of toxins?

Anthropologist Solomon Katz proposes a simple two-pronged explanation: First, it is much easier to prepare food in such a way that the toxins are removed than it is to breed plants with low levels of poison. And second, by allowing the crops to continue making these poisons, we ensure that the plants and fungi are better protected from pests and we consequently increase overall crop yields.

While we humans use elaborate methods to remove plant-based toxins that can harm us, we also spend inordinate resources to acquire *other* potential toxins in the form of spices, which we add to our food and drink. The implication is that spices don't harm us and may actually enhance our health and wellness even though they contain chemicals we know will dissuade and punish pests.

Another facet of this food-as-medicine idea is that our use of spices could be rooted in their ability to prevent and treat infectious diseases in particular, as an extension of the neurotoxin regulation hypothesis. I touched on this possibility earlier in the book, when I discussed how vegetarians in some communities in southern India consume large amounts of spices that contain high levels of salicylates and that these diets were associated with lower rates of colon cancer.

As you probably realize by now, one-off studies like these do little to move the needle beyond confirming our own biases. If it turns out that the spice-health connection is one data point among many showing similar relationships between spice eating and health, there is potential for a bigger pattern to emerge, as it does when we step back from a pointillist painting.

In 1999, biologists Paul Sherman and Jennifer Billing compiled a large data set on the human use of spices. They began by defining what a spice is: a dried plant product used either as a seasoning in food preparation or as a condiment added after the food is served. The two then cataloged the use of forty-three spices found in ninety-three traditional cookbooks from thirty-six countries. The cookbooks represented most of the major linguistic groups and all the continents except Antarctica.

Several of their findings may explain in an ultimate, evolutionary sense why we use spices. First, spices prevent spoilage, a major cause of bacterially mediated food poisoning. Indeed, nearly all of the spices called for in these recipes and used at the levels commonly found in food inhibit bacterial growth. The spices most effective at killing bacteria were also more likely to be used across all countries.

Intriguingly, in recipes that include spices containing chemicals degradable by heat, like the chemicals in parsley and cilantro, the spices were added *after* cooking and not before. Those that did not degrade in heat were added during cooking. Thus, the way we use spices is not random; rather, it is informed by whether the toxins they contain are resilient to, or deactivated by, heat.

Second, spices are more heavily relied on in the tropics than in the higher, cooler latitudes. Sherman and Billing further found a positive relationship between annual temperature and both the proportion of recipes with at least one spice and the number of spices per recipe. More recipes from the tropics called for spice, and for more spices per recipe, than did recipes from temperate regions.

As the temperature of the geographic region goes up, so does the use of anise, basil, bay leaf, cardamom, celery, chilies, cinnamon, cloves, coriander, cumin, garlic, ginger, green pepper, lemongrass, mint, nutmeg, onions, oregano, saffron, and turmeric. As the temperature goes down, the use of dill and parsley goes up. The use of spices that are highly inhibitory to bacterial growth also increases with temperature, and dishes made in warmer countries are more inhibitory to bacteria than are those of cooler countries.

Some evidence Sherman and Billings found for synergy between spices in their microbe-killing potential might explain how we use them in various mélanges. For example, food poisoning from *Clostridium botulinum*, which is the source of Botox—yes, the same toxin that is used in the cosmetic industry to the tune of over seven million doses annually in just the United States as recently as 2018—was historically problematic in the production of sausages in Europe. To prevent the growth of *C. botulinum* and other toxin-producing bacteria in sausages, people began using spice blends. Black pepper, along with cloves, ginger, and nutmeg, constitute the French *quatre épices* widely used to make sausages. All four of these spices contain antibacterial toxins. It is easy to see how the traditional use of these four spices worked synergistically to stave off deadly bacterial growth in the sausages.

As groundbreaking as the Sherman and Billings portrait of our relationship with spices was, a newer study revisited the question in 2021, examining more than 33,750 recipes from seventy different cuisines that used ninety-three spices in total. Although temperature and spice use were still positively correlated going from equator to the poles, spice use also correlated with foodborne illness incidence when averaged *across* regions. The findings suggest that people use more spice when the risk of foodborne illness is higher. These problems tend to be more common in poorer countries in warmer places.

None of these findings are conclusive, given the complexities involved. Still, the Darwinian gastronomy hypothesis — that our liking for spices evolved because spices sometimes prevent natural toxins from harming us — fits well into the broader framework to explain the ultimate reasons we use spices: they are food as medicine.

As you might guess from the observations in chapter 10, children younger than ten and pregnant people tend to avoid spices just as they tend to avoid psychoactive drugs. Both spices and psychoactive drugs contain teratogenic chemicals that harm human development.

Although there are longer-term reasons for avoiding spices and psychoactive drugs, like preventing developmental problems, there are shorter-term reasons that people do use these substances. For example, a little-supported hypothesis is that people developed a liking for spices to cover up the taste of spoiled foods and to facilitate perspiration in hot climates.

One of the simplest hypotheses is that we use spices because we like them — they taste good. Spices introduce new sensations that make us feel good or at least different. They enhance our experience because they are mildly psychoactive — they change our state of mind ever so slightly.

As discussed in an earlier chapter, a report on nutmeg's psychoactive properties at high doses was the beginning of Andrew Weil's research career in complementary and alternative medicine. A more recent article suggests that clinicians need to be aware that abuse of spices does occur, even if it is exceedingly uncommon.

One interesting example of spice abuse involved Malcolm X. While in jail in Charleston, South Carolina, in 1946, Malcolm X related that his cellmates who worked in the kitchen would bring him penny matchboxes of stolen nutmeg that he used to get high. When stirred into water, a single matchbox of nutmeg "had the kick of three or four reefers." Historical interest aside, these cases teach us something important about spices and psychology.

Take the case of D., a fifty-year-old man who arrived at a Hamilton, Ontario, hospital in 2013 with symptoms of what appeared to be severe bipolar disorder. He'd been admitted to the hospital twice during the past two years, and bipolar disorder was the diagnosis on both occasions. This time, however, D. revealed for the first time that he regularly ingested powdered nutmeg, sometimes one tablespoon (fifteen grams) at a time. The spice produced all sorts of mental health problems, including dysphoria, hallucination, depression, and, most troublingly, suicidal ideation.

By day three of hospitalization, all of D.'s psychiatric symptoms resolved completely, and he was treated thereafter without any pharmacological intervention. His diagnosis was changed to "substance-induced mood disorder with psychotic features." In other words, his compulsion to use nutmeg was the root of his mental health crises; nutmeg's effect on his mental state was a symptom rather than the cause of his first diagnosis, bipolar disorder. Like AUD and opioid use disorder, D.'s drug use disorder was not a moral failing but a diagnosable illness.

D.'s case is an extreme example. The vast majority of people who attempt to get high on nutmeg do not become dependent, nor would they even think of it as a substance worthy of abuse, given the terrible side effects. The same is true for the rest of the spices in your pantry. But all spices do contain one or more psychoactive toxin, albeit at low doses. Could the subtle, perhaps even subconscious, uplift or shift in mood or perspective be a proximate cause of our use of spice? Are we just microdosing?

Still, I love to open jars of saffron threads in our spice cabinet and inhale deeply. The threads are the stigmas and styles from certain crocus flowers. The smell brings a smile to my face, and I immediately forget whatever it was that I was preoccupied with. An even more pervasive feeling of satisfaction washes over me when I eat the saffron risotto I like to make. The one-of-a-kind perfumy taste brings me pleasure, but it may not be the flavor alone that's responsible for this feeling.

Saffron is widely used as a Persian folk and alternative medical remedy to treat a variety of illnesses, including depression. Small double-blind, randomized, placebo-controlled clinical trials show that saffron may be as effective at treating depression as the generic form of fluoxetine (Prozac). A number of toxins in saffron, including safranal, which gives the spice its particular odor, may explain saffron's potential effectiveness as an antidepressant.

The most interesting study, to my fanatically analytical scientific mind, was one done on lab rat models of obsessive-compulsive disorder. In this study, the rats were chemically induced to obsessively groom. When given glycosides known as crocins, one of which is transformed into safranal, the rats stopped the obsessive-compulsive-like grooming behavior. Safranal also has anticonvulsant effects in rat models, possibly because it triggers the $GABA_A$ receptors—the same ones that alpha-pinene, betulin, and ethanol all bind to—yielding a hypnotic, calming effect. Safranal is just one example of the innumerable psychoactive toxins found across the diversity of spices.

There is yet another possible explanation for why we like some spices: they cause discomfort, and sometimes, this discomfort can lead to pleasure. In the following section, we examine the link between pain, pleasure, and drugs of abuse.

Spicing Things Up

Relief from physical pain or mental anguish is rewarding in itself. So, when we consume spices like wasabi, chili, and mint, we get to that state of relief only after moving through discomfort. The pleasure we associate with eating these spices may be partly due to the alleviation of the pain it causes. Such a proximate reason for consumption of pain-causing spices—namely, that it eventually feels physically good after we eat them—could go hand in hand with an ultimate, evolutionary benefit, like the suppression of foodborne infection or the general enhancement of health and wellness. By exploring how we sense and respond to spices, we will also learn about spices' connection to the human susceptibility to drug abuse disorders.

Beyond activating our sense of smell and taste, spices also engage another set of receptors in our somatosensory system. When activated, the somatosensory receptors allow us to sense pressure, pain, numbness, tingling, heat, cold, and itch. These receptors are found in nerve cells lining the inside of the mouth and in nerve cells located throughout the rest of the body, from the skin to the gut and even in the brain.

I mentioned several of these receptors already. The "wasabi receptor" TRPA1 detects changes in temperature but also senses noxious chemicals that we perceive as hot or spicy, like the mustard oils in wasabi. The "capsaicin receptor," or TRPV1, is found in all animals with a backbone, and like TRPA1, it is also involved in sensing thermal heat and physical pain in addition to sensing capsaicin from peppers. Finally, the "menthol receptor," or TRPM8, is responsible for pain sensations that arise from cold temperatures. TRPM8 allows us to detect menthol from mint, and methyl salicylate from oil of wintergreen.

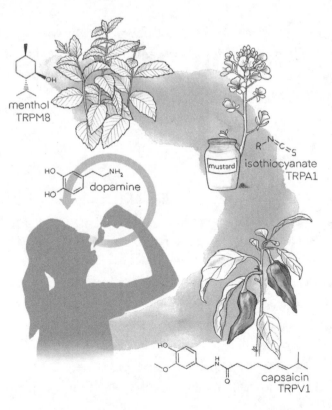

More generally, the TRP receptors, which are found across a diversity of animals, have been targeted repeatedly by plants, fungi, and even other animals like ants that seek to defend themselves using chemicals. In fact, every spice you would find sold in the spice markets along Khari Baoli in Delhi, India, contains toxins that activate at least one TRP receptor.

The long and incomplete list of TRP-activating spices and other ingredients includes the following:

allspice	coriander	nutmeg
anise	cumin	paprika
asafetida	curry leaf	poppy seed
asaron	dill	olive oil
basil	fenugreek	oregano
bay leaf	galangal	parsley
black pepper	garlic	red pepper flakes
cassia	ginger	sage
cayenne pepper	horseradish	saffron
celery seed	lemongrass	sesame seed
chervil	lemon zest	tarragon
chili	mace	thyme
cilantro	mint	turmeric
cinnamon	mustard seed	vanilla
clove	onion	

While these spices and other ingredients activate one or more of the TRP receptors in our mouths (and elsewhere!), they also contain chemicals that activate our odorant and taste receptors in the nose and mouth. Consequently, each of these spices has a distinctive smell and taste, but each also causes us to feel heat, cool, tingling, or numbness (hence somatosensory) in our tongues, noses, and throats by activating the TRP receptors. Through the activation of more than one sensory modality, spices can bring us to a new place psychologically by providing novelty and even distracting us with physical manifestations like sweating. They tend to lift the mood by changing our perspective and activating different chemical senses simultaneously. Often, the pleasurable smell or taste is inseparable from the discomfort, too, so we may learn to positively associate the contrasting sensations, the pleasurable and the not-so-pleasurable.

The same goes for carbonated beverages, which also trigger our TRPA1 receptors. That's why soda stings the backs of our throats while going down and why it burns inside our noses if we burp the carbon dioxide bubbles back out. The taste cells in our tongues detect the carbon dioxide after it is converted into bicarbonate ions and protons in the mouth. The protons, which are acidic, trigger sour receptors called PKD2L1 in our taste buds. Both the sting from the activation of TRPA1 and the sour from the activation of PKD2L1 contribute to the strange appeal of soda. This is an example of how the synergy between our senses can drive dietary preferences. Another source of soda's appeal comes from the sugar and citrus or cola flavor added to many of these drinks. Most of us first try sodas with these appealing additives. Our sweet taste receptors are then activated, and so we learn to associate a positive experience with the uncomfortable feeling of the bubbles, much as we associate the positive effects of the caffeine in coffee with its highly bitter taste. The same goes for spices like capsaicin and wasabi—but for these spices, we need to dig a little deeper to understand what might be going on.

An understanding of why your body reacts the way it does when you touch a hot pan can, paradoxically, help us grasp the human attraction to these sometimes-painful spices. The withdrawal reflex that pulls your hand away from the threat occurs before your brain is even aware of it. In a fraction of a second, heat-detecting nerve endings in your skin fire electrical signals called action potentials to what are called relay nerves deep in your spinal cord.

The neurotransmitter glutamate and a peptide called substance P are two of the chemical messengers our nerve cells use to transfer the pain message in our hand to relay neurons in the spinal cord. The relays in turn synapse with motor neurons, which fire another set of action potentials at a hundred miles per hour through motor nerve cells back to the hand from the spinal cord. The motor neurons then trigger muscles in the hand and arm to involuntarily jerk the limb away from the heat. All of this happens so quickly that it takes us by complete surprise. We are not even aware it is happening until it's over.

Then the terrible, stinging pain of the burn sets in. Pain and inflammation signals travel through nerves to the spinal cord and then to the brain. In response, the pea-sized anterior pituitary gland in the brain releases endorphins into the bloodstream.

The endorphins, our most important homemade opioids, are peptides that make their way to the nerve cells in our burned hand and bind to the opioid receptors there. By doing so, the endorphins block pain-propagating glutamate and substance P from being released. The chemical cascade that started with a burst of glutamate and substance P and led to sharp and then dull pain is dampened by endorphins.

The endorphins released deep in our brains also bind to opioid receptors in the brain itself, not just at the site of the pain. When this happens, the nerve cells in our brain stop releasing the neurotransmitter GABA. GABA, if you recall, is an inhibitory neurotransmitter that dampens our brain's activity by binding to $GABA_A$ receptors, helping us to calm down and fall asleep. As levels of GABA go down, levels of dopamine released into the synapses within our brain's mesolimbic reward center rise, all thanks to the initial release of endorphins by the pituitary gland. Within minutes, the pain in our hand and our concern over it diminishes, thanks to the release of endorphins, followed by a pulse of dopamine.

What does the hot pan example have to do with the pleasurable feeling associated with consuming chilies? When we eat chilies, our reaction is *also* first triggered by the activation of TRPV1. The pain and inflammation caused by the capsaicin binding to this receptor triggers the release of endorphins that block the pain signal in the mouth and cause the dopamine-releasing neurons in the mesolimbic reward system to fire and release dopamine.

Some of us release more dopamine and endorphins than other people do. Moreover, the number of dopamine and opioid receptors in the brain's mesolimbic reward system varies from person to person. People with lower levels of these neurotransmitters and their receptors may be more at risk of developing drug use disorders. These people have less reactive dopamine- and endorphin-mediated brain pathways, which play a critical role in helping a person cope with stress, neglect, fear, and pain. It is easy to see why later in life, survivors of childhood trauma like my father are more susceptible to using drugs of reward that activate these systems just to feel normal.

There are complex reasons why some people have lower levels of dopamine and endorphin and fewer receptors. Both inherited and environmental factors come into play, and these factors interact. A critical piece of this puzzle is the quality of early childhood development.

Addiction medicine specialist Gabor Maté points to four brain systems affected by parenting quality: the endorphin-based attachment-reward system for forming loving bonds, the dopamine-based system that motivates, the self-regulating behavior system based in the prefrontal cortex, and the stress-response system. Each system is controlled by a delicate balance of endorphins, dopamine, serotonin, and norepinephrine. When this balance is disrupted, we often use external stimuli, or drugs of abuse, to compensate.

An important connection to evolution here reveals an Achilles' heel in how our brains develop. The circuits needed to navigate the adult world are laid down early in both humans and other primates. However, we are unique among primates in that most development of the human brain occurs *after* birth rather than before. If a newborn human's head were to contain a brain as large as that of today's two-year-old, the head simply couldn't have made its way through the mother's pelvis. For this reason, humans evolved to shift the brain's growth spurt to the period after birth. So, while the womb shields a developing chimpanzee's brain until it is well developed, much of that same developmental process unfolds in the first few years after a human baby is born.

Infants and young children raised by abusive, neglectful, or stressed parents will tend to have brains more poorly wired to meet the challenges of being an adult. The fragility of young brains is one reason why humans are so susceptible to drug use disorders.

Drug use disorders may seem like a heavy topic for a section on spices, but the subject has an important connection to why we might like spices, particularly those that trigger discomfort through the TRP receptors. Our emotional responses to stress, pain, and reward are linked.

A particular mutation that some people carry in the *OPRM1* gene is associated with less sensitivity to opioid analgesia. This mutation also lowers the production of the mu opioid receptor 1 in the brain, blunting the mesolimbic reward system. Because of these changes, people with the mutation are predisposed to seek enhanced activation of this pathway by using opioids or ethanol.

A small but intriguing study provides a possible connection between this observation about the *OPRM1* gene mutation and spice use. In the study, men with AUD liked spicy foods more than did the men without AUD. This heightened preference for spicy food was in turn associated with the *OPRM1* mutation that is associated with lower endorphin levels and with fewer endorphin receptors.

Do these associations explain why my father and maternal grandmother—both of them with AUD—so generously applied pain-triggering spices to their food? The answer, of course, is impossible to know. Because drinking alcohol desensitizes us to taste in general, this desensitization may explain why people with AUD may prefer spicier foods. Still, the links between the use of spices that trigger pain of some kind, the anticipation and feeling of reward, and the risk of opioid use disorder and AUD are tantalizing.

Making Sense of It

The trouble with searching for immediate and long-term biological reasons for spice use is that in the end, none is wholly satisfactory. As ethnobotanist Gary Nabhan has said, our diet, our culture, and our genes act both independently and interactively to contribute to our use of spices. This melding of the cultural and biological factors offers the most thorough answer to the question of why we use spices.

Paul Freedman, a historian of food, wrote that medieval Europeans were passionate about spices because of the "prestige and versatility of spices, their social and religious overtones, and their mysterious yet attractive origins. Versatility is especially significant because...spices were not used for just cooking. They were regarded as drugs and as disease preventatives in a society so often visited by ghastly epidemics...[T]hey were not only medicinal but luxurious and beautiful."

Part of the early mystique about spices came from the long-held belief that the Garden of Eden, or paradise, existed in Asia. People thought that the plants growing in Eden were endowed with special healing and spiritual powers that Adam and Eve had access to before they were banished from paradise. Today, we primarily think of spices as ingredients or condiments, but for hundreds of years, they meant far more to us. Bound up in legend, they were used in food, medicine, and spiritual practice, not unlike many of the drugs we use for the same purposes now.

If it seems strange that spices would be used as medicines, a predominant theory in medieval Europe passed down from the ancient Greek philosophers held that illness manifested itself when the fluids of the body were imbalanced. Spices were used to rebalance the "humors." Moreover, people used spices of all kinds to prepare theriac, a concoction used as a panacea in many places, including medieval Europe, for over a thousand years. The roots of theriac can be traced to Mithridates VI Eupator, who was born around 135 BCE. Afraid of being poisoned, he took small amounts of toxins in a desire to immunize himself in case somebody attempted the murderous deed.

Out of Mithridates's paranoia the poison antidote mithridate was born. The recipe called for thirty-six ingredients, including opium poppy and viper. Mithridate was then co-opted by the Romans, who called it theriac. In the first century CE, Greek physician Galen popularized a theriac that, he wrote, allowed chickens to survive snake bites, according to experiments he had performed.

Although it may have done nothing more than cause opium dependency, theriac was believed to do much more. Its use spread across Europe and was part of the pharmacopoeia. Painted porcelain theriac jars took prominence in apothecaries and even showcased scenes from antiquity.

Beyond their use to stave off the plague, to rebalance the humors, and as ingredients in panaceas, the food of medieval Europe was spiced to the point of absurdity. For so many reasons, Asian spices seemed to cast a spell over the medieval European psyche. The global historical consequences of this demand for Asian spices are incalculable. But even after spices ran their course, there were other, coveted natural toxins that also caused geopolitical tipping points.

In the next chapter, we will examine how each of four bellicose periods of the last five hundred years hinged on the pursuit of just four alkaloids: myristicin from nutmeg, morphinans from opium, caffeine from tea, and quinine from cinchona.

12.

Nutmeg, Tea, Opium, and Cinchona

*Those narrow straits of Sunda divide Sumatra from Java; and
standing midway in that vast rampart of islands, buttressed by
that bold green promontory, known to seamen as Java Head; they
not a little correspond to the central gateway opening into some vast
walled empire: and considering the inexhaustible wealth of spices,
and silks, and jewels, and gold, and ivory, with which the thousand
islands of that oriental sea are enriched, it seems a significant
provision of nature, that such treasures, by the very formation of
the land, should at least bear the appearance, however ineffectual,
of being guarded from the all-grasping western world.*
— HERMAN MELVILLE, MOBY-DICK; OR, THE WHALE

Silk Roads and Spice Voyages

Spices flowed into medieval Europe, thanks in large part to the Arab traders who sailed from the Horn of Africa to South Asia and back again,
bringing their wares from Red Sea ports overland to Alexandria. From
there, the spices were transported across the Mediterranean to intermedi-
aries in Genoa, Venice, Provence, and Catalonia.

The overland Silk Road also played a role in moving spices and other
Asian imports. Genoese traders established a trading post at the Black Sea
port of Caffa (today's Feodosia) on the Crimean Peninsula in 1266, with
the blessing of the ruling Batu Khan of the Golden Horde. That agreement
ended when the Mongols sieged the city in 1343. But then *Yersinia pestis*, the

bacterium that causes bubonic plague, interrupted their plans—by 1346 the Mongol army was overrun by the Black Death.

The plague had hit Europe before, but not this hard. In the bodies of fleas, rats, and, most importantly, people who had escaped the siege in Caffa, the disease moved ashore in 1347, when they sailed into the port of Constantinople. The Black Death then struck mainland Europe around 1348.

In just three years, the plague killed off half of Constantinople's population. The decimation set into motion its fall to the Ottoman Empire a century later, in 1453. As the city fell, so did the 1,400-year-long reign of the Roman Empire.

With no knowledge of the germ theory of disease, people thought that "corrupt air" was the cause. Like the perfumed gloves of Catherine de' Medici, perfumes and pomanders of aromatic herbs, resins, and spices were heavily relied on to keep the plague at bay. Along with its use as one ingredient in many concoctions of theriac, nutmeg itself was used as a panacea and became a particularly important ingredient in plague treatments.

Demand for Asian spices was high as the Black Death unfolded, but the Ottomans had taken control of Venetian territory, from the Adriatic to Egypt and the Levant, and, with it, the European spice trade. Could this explain why the medieval kingdoms of Europe began to seek their own sea routes to Asia?

It may have been one of the factors, but the Ottomans continued to supply Europe with spices. It was not only the fall of Constantinople to the Ottomans or any other crisis with the Muslim world per se that drove the medieval European kingdoms to seek their own sea routes to Asia to gain access to spices. Rather, says Freedman, the "discovery of a water route to the Indies was...more of an opportunity than a necessity."

The opportunity was in part born through the accumulation of geographic knowledge, including information gained from Marco Polo's expedition across the Silk Road coupled with ever more daring seafaring excursions along the African coast. The countries positioned closest to the West African coastline, Portugal and Spain, were the first to explore it.

Christopher Columbus persuaded Ferdinand and Isabella of Spain to fund an expedition to find the East Indies via a western route. Believing he had reached Asia, Columbus landed in what is now the Bahamas on October 12, 1492. Although he didn't know it, the trip was far more impactful than it would have been had he actually landed in the East Indies as intended.

Columbus sailed to a continent completely unknown to Europeans, except for the explorations and short-lived settlements of the Norse. None of the familiar Asian spices were found on Columbus's voyage. Columbus and his men captured several Indigenous people and returned to Spain with them in 1493, along with animals and plants—the beginning of the Columbian exchange.

Meanwhile the *São Gabriel,* Portuguese explorer Vasco da Gama's carrack, rounded the Cape of Good Hope and sailed to India, reaching the Malabar Coast on May 20, 1498. One man was sent ashore first, whereupon he declared that they had come in "search of Christians and spices."

Not long after da Gama's voyage, the Portuguese came to control much of the Malabar Coast by taking advantage of rivalries across its port cities. A Portuguese near monopoly over black pepper and ginger was at hand, and factories were built by Viceroy Francisco de Almeida in King Manuel's Estado de India by 1505. By 1506, de Almeida's son Lourenço led a fleet to Sri Lanka, the principal source of cinnamon, and in 1518 the Portuguese established a factory to process spices there in Colombo. Like the black pepper and ginger bonanza flowing from the factories in India, this development lifted the constraints on cinnamon shipments to Europe.

In 1511, the Portuguese unceremoniously reached the Spice Islands, or the Moluccas, in what is now Indonesia. Although Europeans' arrival in the Moluccas did not seem like a historically remarkable event, early colonization became a geopolitical turning point. The trees that produced the three most coveted spices—cloves, mace, and nutmeg—were only known from these islands. Because the Portuguese had colonized most of these islands, Portugal controlled the early spice trade for nearly one hundred more years.

It was a short but profitable run. By 1579, the Portuguese began to lose their grip. That year, Francis Drake's ship, the *Golden Hind*, made its way to the Moluccas. Drake brokered an alliance between Queen Elizabeth I and the sultan of Ternate, but the next two voyages under Elizabeth I failed. These botched efforts created an opening for the Dutch to challenge the Portuguese for control of the Spice Islands. In 1595, ships of the Compagnie van Verre, the forerunner of the Dutch East India Company, set sail with that objective.

Meanwhile, the English East India Company was chartered in 1600 to make yet another push for a slice of the spice pie. In 1603, four British ships returned to London along with around one million pounds of black pepper. The ensuing glut caused the price of pepper in Europe to drop.

The Dutch focused on seizing control of clove, mace, and nutmeg production, which remained highly valuable in Europe. By 1601, over fifty Dutch ships had reached the Far East, and factories were set up in the Moluccas to process the spices. There were still pockets of resistance, including a group on Run, one of the Banda Islands still claimed by the British.

In 1667, the Treaty of Breda stipulated that British-claimed Run would become Dutch territory, while the Dutch-claimed Nieuw Nederland (the region running roughly from what is now Delaware to Cape Cod) would be transferred to the British, among other concessions. Of course, New Netherland included the island of Manhattan. To say that Run was traded for Manhattan because of a war over nutmeg would be hyperbolic but not entirely unfounded. Among other things, control over nutmeg, a substance that by the seventeenth century had come to be seen as a panacea, including a cure for the plague, was part of the equation.

The mystique of myristicin, the principal alkaloid in nutmeg, still held plenty of sway over the affairs of Europe as the continent entered the modern era. However, as demand for spices began to wane, two other alkaloids, morphine and caffeine, helped foment the fall of the Qing dynasty and the rise of modern China.

Tea for Opium and Run for Manhattan

The Treaty of Breda—like Portuguese control over the TRPV1-triggering black pepper and ginger, and the "discovery" of the Spice Islands themselves—was a watershed moment in global history. After signing the treaty, the Dutch focused on Indonesia and spices. The British looked to India and China for their pursuits, including the opium trade. They used India as a place to grow poppies and used China to sell it—or, more precisely, to smuggle it.

In the seventeenth century, European cuisine turned away from spiced food, and medicine turned away from the previously mentioned humoral theory, the belief that diseases were caused by an imbalance of humors. But as spice demand waned, demand for other more psychoactive plant products began to take hold through the Columbian exchange and the overwater trade routes from Asia. Stimulants included nicotine from American tobacco as well as the caffeine and related methylxanthine alkaloids from South American chocolate, African coffee, and Chinese tea.

As historian Stephen Platt notes, the first British person ever to sip a cup of tea from the leaves of the evergreen tea shrub *Camellia sinensis* was probably Captain John Weddell, who in 1637 commanded a ship for King Charles I, with the goal of buying ginger and sugar from China. Although his trip didn't quite go as planned, tea soon displaced coffee as the preferred hot beverage in Europe. Tea emerged as a cure-all and font for wellness.

By the 1660s, Dutch and British physicians were extolling tea as a panacea. Although the British imported small amounts from the Dutch, the beverage became so popular that the nation wanted a direct supply. In 1717, direct trade with China was established when the East India Company shipped tea from Canton to Britain for the first time. The tonnage imported by Britain grew with each subsequent year, from approximately 250,000 pounds in 1725 to 24 million pounds in 1805.

Things were initially going well on the British side. Textiles from Britain were coveted by the Chinese. But by 1759, Emperor Qianlong forbade the British from trading in cities beyond Canton. The British, French, and Americans were suddenly restricted to doing business in trading centers (factories) set within a walled-off plot of reclaimed land in Canton. These restrictions proved to be a sticking point as European lust for Chinese markets grew. Pushed into a corner, the British began to pursue other ways to rebalance the trade deficit created by the demand for Chinese tea. They turned to the opium grown in India and began to smuggle it into China to sell.

Platt's analysis of eighteenth-century Chinese-European relations reveals that the Chinese Empire was the most prosperous on the planet by any measure, from its standard of living to the size of its population, which accounted for one-third of humanity at the time. The Chinese, along with the Europeans and Americans, all recognized China's top status. And China's arm's-length imperial trade policy was one practice of many designed to protect its status. There was a reason the emperor forbade foreigners from learning Chinese.

However, the corrosive effects of political corruption, the growing needs of a population of more than four hundred million, and the rise of a religious, millenarian sect in the hinterlands called the White Lotus all converged to threaten social and economic stability in late eighteenth-century China. The White Lotus Rebellion was violent and influential, and the empire's attempts to quash it proved financially ruinous. His power weakened, Qianlong abdicated the throne in 1796 and died three years later.

Although opium was officially banned in China at the time of Qianlong's death, the drug was widely used and viewed as a luxury item when smoked in combination with tobacco. The misery index in China began to rise in parallel with the rise in opium use. And as we have learned, it is among the vulnerable, the stressed, and the victimized that drug use disorders are most likely to take root.

By 1828, half of the value of British goods sold in China was in illicit opium, outpacing even the value of all the tea imported to Britain that year. The dominance of this plant-based trade was another global tipping point, but instead of one caused by spices, this one hinged on the value of the morphine in opium and caffeine in tea. As mentioned earlier, the British became so focused on opium because the profits from the vast amounts smuggled into China from India allowed the British to rebalance an ever-deepening trade deficit over the insatiable British demand for tea from China.

The caffeine in tea was never going to bring down the British Empire, but the morphine in opium just might have hastened the fall of the weakened Chinese emperor Qianlong. Though the prohibition against its importation and use was technically in force, by the mid-1830s, opium was so heavily used throughout China that even the heir to the dynasty, Prince Mianning, smoked opium.

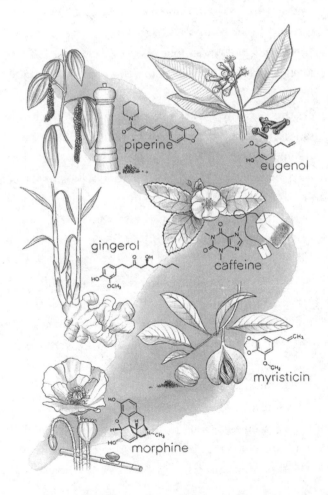

Despite considering opium legalization in 1837, the Chinese government, at the behest of the emperor, doubled down on enforcing the ban, formally cracking down on individual users, dealers, and, finally, the source of illicit opium: smugglers in Canton. In 1839, the factories were shut down and hundreds of foreigners were imprisoned on the demand that all opium be turned over to the authorities, who eventually dumped thousands of chests of the drug into the river. All foreign trade with China stopped at the mouth of the Pearl River, and opium use in China became punishable by death.

Of course, the British opium traders still wanted to be compensated for the opium the Chinese authorities had destroyed. To settle the matter, the British sent a small fleet of warships to blockade the ports and force China to surrender. But before they arrived, China banned Macau from supplying British trade ships, and a skirmish broke out between them and Chinese war junks in September 1839. Thus began the first battle of the Opium Wars.

The British Parliament then authorized war, and by the summer of 1840, a large British fleet had taken control of the Pearl River, of Canton, and of the surrounding area. In negotiations carried out to keep the peace, China agreed to pay Britain three times the value of all the opium that had been dumped into the Pearl River and to designate Hong Kong as a factory, but Britain was granted no new ports of call.

Neither country signed the treaty and the Opium War continued, with nearly every battle won by the British. The conflict ended in August 1842, when the British threatened to sack Nanjing. The Treaty of Nanjing, signed by Queen Victoria and Emperor Daoguang, was about as lopsided as one could imagine. The Chinese would pay Britain fully ten times the cost of the dumped opium, Hong Kong would become permanent British territory, and several other cities, including Shanghai, would open up to trade. So began the Chinese "century of humiliation."

Opium dependence may have ensnared upwards of 30 percent of the Chinese population by the end of the nineteenth century. And as the Qing dynasty fell after the Xinhai Revolution ended in 1912, the country entered a period of civil war, invasion, revolution, and other unrest. Opium use continued to be widespread until Mao Zedong outlawed it in 1949 and forced ten million people suspected of having what we now call opioid use disorder into treatment centers.

Although internal strife associated with the White Lotus Rebellion and widespread corruption had weakened the Chinese Empire internally even before the opening battle of the Opium War, historians agree that the cheap opium flowing in from British-controlled India contributed to its decline. One perspective is that the British Empire's rapacious desire for economic gain and hegemony led to the downfall of the most advanced civilizations the world had known.

By the latter half of the twentieth century, icy relations with the West had begun to thaw, and in 1997, the United Kingdom returned Hong Kong to China. China's reascent continued as the opioid crisis unfurled in the United States. After a devastating "century of humiliation," China reclaimed its preeminence on the world stage.

The waves continue to ripple. As average life expectancy in China increases year after year, it has begun to decline in the United States. One cause is the US opioid epidemic, including a wave propagated by illicit fentanyl.

In an ironic twist, China has been the main source of fentanyl trafficked in the United States. Under mounting pressure, the Chinese authorities restricted the use of precursor chemicals to make the drug. Piperidine is the basis for two such precursors and, as described earlier, is also the building block of many other alkaloids.

After the restrictions took hold in China, the production of precursors, illicit fentanyl, and fentanyl-containing pills has begun to shift to India and Mexico. It seems that the supply chain has simply diversified. Just as the widely available British opium that flooded the market in China does not fully explain the fall of the Chinese Empire in the early nineteenth century, the semisynthetic and synthetic opioids that flooded the United States both legally and illegally were not a singular cause of political and socioeconomic strife in the United States. Yet at the same time, opium in China and opioids in the United States accompanied dramatic social upheaval and political change and in some ways catalyzed them.

As Donald Trump campaigned for the US presidency in 2016, he promised to end the opioid crisis if elected. This strategy of focusing on poorer, whiter, and more rural voters included the moniker the "forgotten people of America." This demographic was heavily affected by the opioid epidemic, and the campaign message was that the elites in Washington, D.C., just didn't care about them, as proved by the raging crisis.

Although the Trump administration took meaningful steps to address the opioid epidemic in 2018, the execution of those steps was plagued by leadership blunders. Following a small dip in deaths in 2018, American opioid overdose deaths rose again and have done so every year since.

A landmark study in 2015 found that from 1999 to 2013, mortality rates from drug poisonings (overdoses), suicide, and chronic liver cirrhosis for "US White non-Hispanics" rose across every five-year cohort from ages 30 to 64. So strong was the impact of these factors that life expectancy dropped across the fifteen-year interval for the 45- to 54-year-old demographic. The impact was highest in the South and West and among those without college degrees. These findings comported with the more intensive marketing of prescription opioids in whiter, more rural, and poorer locales in the United States. After 2013, however, as fentanyl took over, higher rates of opioid overdose deaths were observed for Black, Latinx, and Indigenous Americans.

After 2010, life expectancy in the United States has stalled overall, largely due to a flatlining of cardiovascular disease risk. Among all causes of death, cardiovascular diseases play a far larger role than do any of the drug-related deaths. Nevertheless, cardiovascular-related deaths may in fact be related to drug-related deaths. Cardiovascular risk greatly increases with opioid use disorder or AUD. So-called deaths of despair include many people who have succumbed to the interactions of these variables.

A case study is my father, who died in 2017 of atherosclerotic coronary artery disease and cirrhosis of the liver, both chronic physical manifestations of his AUD. The dominoes that fell along the way included the loss of much of the value in his already-meager 401(k) retirement funds in the aftermath of the Great Recession, the loss of his short-lived sobriety, the loss of employment, the loss of his marriage, the loss of primary medical care, the loss of immediate family ties, the loss of his brother (and primary support system), and the loss of cognitive function. My father is just one data point in a descending life expectancy trend that reflects the structural inequities in the US social and economic fabric. The United States is alone among "developed" nations in the decline in overall life expectancy, revealing that the problem is of its own making, including the many waves of the opioid crisis.

The ruinous reach of the spice trade and the early modern era extends right across the last five hundred years, from the demise of a two-millennium-long empire to an ongoing public health crisis in the United States. The lust for spices, then caffeine, and then opioids — substances first forged in the war of nature as defensive shields — was pivotal to each of these tipping points, societal and personal.

The other side of the coin is that natural toxins like morphine have also done more to mitigate human suffering and enrich our lives than perhaps any other set of tools at our disposal. Another two-sided story is that of the antimalarial drug quinine, an alkaloid from the bark of the cinchona. As we'll see in the next section, although quinine turned the tide of the Pacific theater in World War II, an underappreciated consequence was the realignment of Central American and South American economic and political ties from Europe toward the United States.

Quinine, War, and Peace

Like the isoprene polymers of rubber, the alkaloid quinine from the bark of the cinchona, or fever tree (*Cinchona officinalis*), was a decisive factor in the outcome of World War II. The two plants that produced commercial rubber and quinine, one in the spurge family and one in the coffee family, were both native to South America. First, the Spanish imported quinine in cinchona bark to Europe. As mentioned earlier, the quinine cured malaria, a disease widespread in Europe at the time of the conquest of the Americas. In the seventeenth century, the Spanish Crown controlled a monopoly on cinchona bark in Latin America for nearly forty years. Although only

350,000 pounds were extracted at the time, it was enough to help the Spanish keep malaria at bay both at home and abroad, and helped Spain achieve its remaining imperialist goals in the New World. The national tree of Peru, *C. officinalis,* even appears in the country's coat of arms, thanks to Simón Bolívar, and the quinine and related alkaloids in its bark are used to make angostura bitters, an ingredient in the national drink, the pisco sour.

Although the Spanish introduced both quinine and rubber products to Europe, it was the British and Dutch who brought the rubber trees and the original cinchona trees to their colonies in Southeast Asia. The goal was to control the flow of each resource through global monopolies to support the colonial pursuits of each country in the Global South. Some historians have interpreted the covert removal of these two trees from South America as secret acts of biopiracy.

After Japan invaded the Dutch East Indies in 1942, it took control of these cinchona plantations. These fever trees, which were native to the Andean cloud forests but were now being cultivated on the other side of the planet in Java, supplied over 95 percent of the world's quinine at the time. Making matters worse for the Allies, by 1940 the Germans had seized control of the largest stockpiles of refined quinine in Amsterdam (imported from Java). The Dutch monopoly on quinine had suddenly become an Axis monopoly on quinine. The Allies simply hadn't planned for such a confluence of events: a frozen quinine supply chain and the greatest need for this drug at the most critical period in the war.

In 1942 alone, more than eighty-five hundred Allied troops fell ill with malaria in the Pacific theater of World War II, far more than had been wounded by the Japanese in battle. On the surface, cases of malaria should not have been a problem, given the availability of Atabrine (quinacrine or mepacrine), the newly synthesized antimalarial based on quinine's molecular structure. The new medication worked to cure malaria, but because of its terrible side effects and a rumor that it caused impotence, Allied soldiers generally refused it.

More quinine was needed if the Allies were to win the war. A push was made to find new sources across the native ranges of cinchona trees in Central and South America. One goal was to create in these ranges new plantations that would be controlled by the United States and its allies. A group of botanists from the University of Michigan proposed a botanical expedition to the US Board of Economic Warfare. The board approved and brought in the Department of Agriculture and the National Arboretum to advise.

Through the Cinchona Agreements, the US botanists on the Cinchona Mission would search for, procure, and export bark and quinine extract from Bolivia, Colombia, Ecuador, Peru, and Venezuela. The parties also hoped to augment pharmaceutical company Merck's decade-old plan to create plantations in Guatemala and extend those efforts to Costa Rica.

From 1941 to 1945, the United States imported a staggering thirty-four million pounds of bark from cinchona trees and around forty-four thousand pounds of cinchona alkaloids from South America. The cache of imported bark grew to more than forty million pounds by 1947. The quinine obtained from the Cinchona Missions allowed the Allies to win the war in the Pacific theater.

Perhaps more cryptically but arguably as important, the missions also fundamentally shifted Latin American loyalties away from Europe and toward the United States. This realignment of interests has had major economic, environmental, political, and social consequences.

But at the beginning of the twentieth century, Latin America's loyalties to its large northern neighbor were not so strong. Spurred by the Monroe Doctrine's imperial subtext, the United States invaded Cuba, Haiti, the Dominican Republic, Mexico (several times), Nicaragua, and Panama in the four-year interval from 1914 to 1918. These excursions ended in 1933, when Franklin Delano Roosevelt announced his Good Neighbor Policy, which held until the end of World War II and the beginning of the Cold War in 1945. The policy took a noninterventionist approach based on the principle of mutual respect. It was designed to strengthen the social and economic bonds between the United States and Latin America.

The Good Neighbor Policy aimed to promote the development of Latin America and, in so doing, obtain the raw materials the United States needed, like quinine, rubber, and timber. From the US perspective, the Cinchona Missions were part of this Good Neighbor Policy—the United

States needed quinine, and the nation could offer incentives in return. The offerings made in exchange for these raw materials—from loans to scientific expertise—had strings attached that coincided with US style of agriculture: large-scale monocultures. In the end, advice from the United States superseded local knowledge. Unfortunately, the adopted US approach involved questionable agricultural and forestry practices like the introduction of unnecessary crops and the subjugation of Indigenous peoples. These practices continue to this day and have hastened the destruction of the world's largest tropical rain forest, among other negative effects. And importantly, many of the remaining gray-barked *Cinchona* tree species of the Andean cloud forests are now threatened, owing to centuries of exploitation followed by deforestation and habitat degradation.

The Cinchona Missions were one small part of a broader US foreign policy to ensure dominance over the economic, social, and political affairs of Latin America. As the Cold War progressed, the United States saw leftist movements in Latin America as an existential threat. Having just extracted tens of millions of pounds of cinchona bark from South America to solve one of its problems, the United States decided that the threat of communism there was actually the source of another problem.

The story of nutmeg, tea, opium, and cinchona reveals how our pursuit of natural toxins altered the course of human history in profound ways. This history is now our reality. Two of the unintended consequences of the obsession and need for natural toxins are the global biodiversity crisis and the global climate crisis. These twin problems threaten not only our survival as a species but also that of the biosphere itself. Ishmael's "all-grasping western world" in *Moby-Dick* may soon be grasping at straws if the trends continue. There is hope for a redemption, but it requires empowering the Indigenous and local communities that live on the lands most affected by biodiversity loss and climate change.

13.

The Future Pharmacopoeia

> There is grandeur in this view of life, with its several pow-
> ers, having been originally breathed into a few forms or
> into one; and that, whilst this planet has gone cycling on
> according to the fixed law of gravity, from so simple a
> beginning endless forms most beautiful and most wonder-
> ful have been, and are being, evolved.
>
> —CHARLES DARWIN, *THE ORIGIN OF SPECIES*

Poison Gardens

The main thesis of this book is that nature's pharmacopoeia didn't evolve
for our benefit. Rather, many of the chemicals we rely on for many of our
medicines and for food, drink, and recreational and spiritual practices
came from organisms that produced these chemicals through evolution for
the organism's own benefit, be it protective or reproductive. Diverse groups
of animals and every human culture have co-opted these chemicals, largely
from plants and fungi. In doing so, our evolutionary and cultural trajecto-
ries have changed as a species, and the fates of each of our individual lives
hinge on these chemicals, for better and for worse.

To thrive, humans have always required access to nature's toxins, and
our descendants will need them, too. Most of these natural toxins are found
in the tropics. Why tropical latitudes produce so many of these chemicals
should come as no surprise to you by now. There are more species squeezed

into the lands and shallow seas between the Tropic of Cancer and Tropic of Capricorn than live in all other latitudes combined. In the midst of this warm, relatively stable climate, innumerable battles between species rage year-round. The chemicals produced on these battlefields are the most diverse and abundant of any region in the world.

Scientists have shown that primary tropical rain forest ecosystems (with no known history of clear-cutting) hold more plant species and, therefore, a greater diversity of natural toxins than does any other tropical habitat. Yet, the plants most familiar to humans are those that grow in our immediate, disturbed environment and those that we clear, weed, plant, and tend in our gardens.

If you or a loved one has had cancer, you might know about Madagascar periwinkle or rosy periwinkle, a flowering plant that is endemic to Malagasy rain forests and produces the alkaloids vincristine and vinblastine. Vincristine chemotherapy targets white blood cells and has boosted the odds of surviving childhood acute lymphoblastic leukemia and non-Hodgkin's lymphoma from 10 to 95 percent. Equally valuable, vinblastine chemotherapy kills all rapidly dividing cells and is now used to treat a variety of cancers, including breast cancers, melanomas, non-small-cell lung cancers, and testicular cancers.

Remarkably, we only have access to these important medicines because the Madagascar periwinkle had already been long used in traditional medicine by Indigenous Malagasy people and local peoples throughout the world. Most likely because of its medicinal properties, the plant was moved across the globe over the past several thousand years. The diseases it purportedly treated were many and included cancer and diabetes.

Eli Lilly researcher Gordon Svoboda included the Madagascar periwinkle in a study aiming to identify promising drugs from plants to treat diabetes because he had heard that this plant was used in the Philippines for that purpose. Independently, Robert Noble and colleagues at the University of Western Ontario were also screening plants used in traditional medicine for diabetes drugs. Noble's team focused on a report from Jamaica describing a "West Indian shrub" from which a tea was made and used locally to treat diabetes. The two teams learned of each other's work at a conference and collaborated. Unfortunately, however, the plant extracts failed to lower blood sugar in animal models.

While Noble and his team were further studying the rats they had injected with Madagascar periwinkle extract, they serendipitously discovered that the mice had died from infections caused by very low white blood cell counts. The periwinkle extract killed white blood cells. Thus, by studying in detail the pharmacological properties of a tropical plant used in traditional healing, the researchers had discovered powerful new drugs to treat cancers caused by white blood cell proliferation.

The Madagascar periwinkle, often used as an exemplar for the preservation of pristine tropical rain forests, is one such weedy species found throughout the global tropics, although it is now endangered in its native Madagascar. You have probably seen related *Vinca* species cultivated as ornamentals; you may have even thinned these viny but pretty weeds in your own landscaping as I have. So, yes, it is a tropical plant, but it is a weedy one that has now spread across the planet. Humans became aware of the plant because they lived near it and began to use it, and then this knowledge spread to several continents well before the Madagascar periwinkle entered the modern medical literature.

Many important drugs come from weedy species that thrive in disturbed tropical habitats, including those created by subsistence farmers. These plants' ignoble origins don't devalue the inherent or practical value of intact, primary tropical rain forests. But the variety of habitats—both untouched by humans and otherwise—that yield plants of importance to us reveals a more complex reality.

The origin of our modern pharmacopoeia cannot be understood without an appreciation of the dynamic land use practices of Indigenous and local people who have lived in these ecosystems for millennia. In all the tropical biomes of the world, humans have been present and augmenting the landscape for hundreds to tens of thousands of years. The Mayan ruins, for example, rising from what is now dense jungle in Mexico give testament to this long heritage of cultivation. So do the eleven recently discovered 1,500-year-old human settlements under a rain forest in the Bolivian Amazon. The notion that these regions are pristine, or untouched by human influence, is inaccurate and rooted in the same tropes that motivated colonial and imperial conquest for centuries. Today's primary tropical rain forest may have been a village at some distant point in the past. To varying degrees, tropical ecosystems have felt the impact of humans for as long as humans have been living in them. The most diverse ecosystems are not prehistoric time capsules.

Both the mystique of the tropics and the value of its natural products drove the chain of events that led to our modern geopolitical order. So many aspects of our industrialized way of life—the tires on our cars, the drugs that save our lives, the spices we grind every day—depend on the products of ecological interactions between species in the tropics. Many of these natural products that we take for granted were first tapped by Indigenous peoples, largely in the tropics, from species living both in the primary rain forest and at its edges, in gardens. Yet the knowledge and sovereign lands of Indigenous peoples hold much more than just the future of the pharmacopoeia.

Our past actions have finally caught up with us. The intertwined crises of our era—the biodiversity crisis and the climate crisis—now threaten the survival of the most diverse Indigenous cultures and biomes on the planet.

The Apportioner

As we touched down at Gatwick Airport, the roar of the 747's four great engines thrusting in reverse was drowned out by the beating drum of my heart. Finally, at the age of fifteen, I had escaped the gravity of Minnesota at least temporarily, and doing so was thrilling.

I'd never been to another country before, not even Canada. The euphoria of being dropped into the city of London was incomparable. It was the first baby step in my escape from the jaws of a closeted, rural life I didn't want.

It was also the beginning of a long Minnesota goodbye. I knew I would miss the great gray owls and their egg-yolk eyes, the frosty pepper-up-my-nose feeling from the alpha-pinene-infused air of the balsam fir, and the howling of the timber wolves through the frigid air. But those wild things would remain for others to see, smell, and hear.

As we toured museums, cathedrals, and châteaus, I was fascinated by the tapestries hanging on the walls. The tapestry that intrigued me the most hung in a dimly lit room at the Victoria and Albert Museum in Kensington. Of Flemish origin, it was woven sometime between 1510 and 1520, just as Europe began its brutal, imperial conquest of the planet in pursuit of spices and power. The tapestry depicts the Triumph of Death over Chastity, a sonnet in a fourteenth-century Tuscan poem by Petrarch. Three goddess sisters—the three Fates—stand tall over a woman whose lifeless body is enchained.

The scene is framed in a millefleur design of colorful flowering plants and animals in a meadow. I recognized some of the species. Tobacco, strawberry, and a long white lily cut from its base, alongside the fallen Chastity—representing the Virgin Mary.

The three Fates, or three Moirai, from the ancient Greek pantheon, were the weavers of human destiny, the apportioners of life. As a newborn baby took its first breath, Clotho, the youngest sister, spun out raw fibers into the thread of life from her distaff. The thread was passed to Lachesis, the allotter, whose rod took the measure of life. Finally, the oldest sister, Atropos, the inexorable, cut the thread at the prescribed time of death. *Atropa belladonna* (deadly nightshade) was named after her.

I wondered why the three Fates myth from antiquity was worthy of such an ornate tapestry. It became clear to me that, like so many pieces of great art, the tapestry illustrates how death visits us all. The weaver captured this truth. The rabbit's woven eye in the corner tracked me down the hallway, its silken threads reflecting the light of a candle long extinguished. I thought of my own life, where it might go, and when it might end.

Although this tapestry resonated in the way it was perhaps intended, I also felt, and still feel today, that our society, our species, is the apportioner.

But instead of the thread of each human life, all life on the planet as we know it is at stake. We now have powers simply unimaginable five hundred years ago. We hold the power to self-destruct, to cut the thread of all of life and apportion our own fate along with it.

I am not a doomer. I have hope and faith in a green redemption that lifts all boats. We are the one species that we know is capable of transcendent consciousness—a capability that enabled us to leave the planet and land on the moon. We also stand alone in our ability to use our big brains to work together and fix the two most important global environmental problems we have caused: the biodiversity crisis and the climate crisis.

Once we lose a species, it is gone forever. Yes, new life resprouts from the graveyard of mass extinction. Old ecological roles are eventually refilled as the few surviving branches grow and split evolutionarily. But it took tens of millions of years to approach the rich complexity that preceded the five cataclysmic mass extinctions of the past. If we cause another mass extinction, then biodiversity may one day return, but we won't be here to watch the new garden grow.

The tropics hold most of the planet's biodiversity, and they are enormous carbon sinks as well. Although Indigenous people represent 5 percent of the global human population, 80 percent of the world's biodiversity is found in their lands. Much of their forests, coral reefs, peatlands, savannas, and grasslands teem with life and capture an enormous fraction of global carbon dioxide. Nearly one-half (45 percent) of the intact forests of the Amazon basin are Indigenous lands, for example. That is four million square kilometers. So, whether your descendants will inhabit a planet still teeming with "endless forms most beautiful and wonderful" and one that continues to provide the air we breathe, the food we eat, and the toxins we use depends on the support and protection of Indigenous rights and sovereignty everywhere.

There is no Planet B for us, at least for now, and we are still in what should be the early years of our life span as a species. *Homo erectus* was around for two million years before we evolved, and we've been around for only two hundred thousand. Whether we die out as a species and take much of the biodiversity of world with us as we go is in our own hands.

In *Braiding Sweetgrass,* Robin Wall Kimmerer, environmental scientist, writer, and member of the Citizen Potawatomi Nation, writes of the Shkita-gen, the People of the Seventh Fire. An Anishinaabe prophecy holds that after a generation of ruin, another will be born that must "rekindle the flames of the sacred fire, to begin the rebirth of a nation." Let us hope they are among us now.

Acknowledgments

It is all but impossible to thank all the people who helped bring this book to fruition. Still, a few stand out.

I owe a debt of gratitude to my husband, Shane Downing. He has been my rock and my stay. Shane was there at the start, the end, and the in-between. He was there the morning I found out I was awarded a Guggenheim Fellowship to write this book, and he was there late at night when I turned in the final draft more than two years later. The ring he put on my finger in the middle of it all was filled with nature's toxins. It marked a new chapter in the book and my life.

My mother, Layne Whiteman, and brother, Seth Whiteman, gave me their blessing to share some of the most difficult aspects of our lives with you. They supported my view that sharing our story could help serve the greater good. The same gratitude applies to my extended family. My cousins Kelly Johnson, Paige Mellinger, and Rebecca Levenson provided key support.

I sincerely thank my agent, Russell Weinberger, and editors Tracy Behar and Ian Straus. They helped guide my ideas, sculpt the story, and, most importantly, reined me in. Their commitment to the way I thought the story of nature's toxins needed to be told was unwavering. Copyeditor Patricia Boyd helped refine the text in a thoughtful, expert way.

Julie Johnson at Life Science Studios, the book's talented illustrator, was a true creative partner as I wrote the book. The field-notebook style of her drawings perfectly captured the essence of how I see nature and the dialectic between the toxins, the organisms, and us. A picture is worth a thousand words, and Julie's work certainly is so, with her simple drawings to communicate complex concepts. Ecological chemist Christophe Duplais generously checked the accuracy of the chemical structures in these drawings.

I am grateful to Karin Fyhrie for her help in conceptualizing the book's cover.

Countless conversations with colleagues, mentors, and friends over the years helped influence the book. Jim Poff, Bob Sites, and Patty Parker helped get my career off the ground, and I'm grateful to them. Many years ago, Naomi Pierce and I wrote a review article that used "delicious poison" in the title. Both her mentorship and Fred Ausubel's put me on a steady course as a postdoc to study plant-insect interactions as a career. Anurag Agrawal's and Jennifer Thaler's support also proved critical to my research trajectory and in the writing of the book. May Berenbaum, Ian Billick, Mike Botchan, Lynne Cadigan, Nicole King, Kailen Mooney, Corrie Moreau, Michael Nachman, Peter and Cindy Reinthal, John Rahm, Neil Shubin, Cassie Stoddard, and Peter Raven provided early encouragement. I am grateful to Joy Bergelson, Erica (Bree) Rosenblum, Nancy Moran, Kevin Padian, and Michael Silver for expert feedback on the manuscript. Elizabeth (Liz) Bernays gave me detailed comments, line by line, on every chapter. As a major contributor to the field of plant-animal chemical ecology and evolution, Liz gave me advice that improved the book beyond measure.

I thank the John Simon Guggenheim Memorial Foundation for awarding me a fellowship and believing in the story of nature's toxins. The Whiteley Center at the University of Washington's Friday Harbor Laboratories and the Rocky Mountain Biological Laboratory hosted me during key phases of the writing process. The National Institutes of Health supported my basic research, which I discuss in the book: plant-animal interactions mediated by toxins important to human health. My terrific colleagues at the University of California, Berkeley, have been incredibly supportive of me throughout this project. In that vein, I acknowledge all the undergraduate students, graduate students, postdoctoral scholars, and research specialists whom I have been lucky to work with and mentor, from whom I have learned so much, and in whom I have great faith.

Endnotes and an appendix for this book can be found at:

https://www.mostdeliciouspoison.com/notes.html

Index

About the Author

Noah Whiteman, PhD, is an evolutionary biologist at the University of California, Berkeley, where he is Professor of Integrative Biology and Molecular and Cell Biology. He holds affiliations at Berkeley with the Helen Wills Neuroscience Institute, the Center for Computational Biology, the Museum of Vertebrate Zoology, the Jepson and University Herbaria, and the Essig Museum of Entomology. Whiteman is also a principal investigator at the Rocky Mountain Biological Laboratory in Gothic, Colorado. He has received recognition for his teaching and scholarship, including the Harvard University Distinction in Teaching Award, and has been elected to the Royal Entomological Society, California Academy of Sciences, and Board of Directors of the Genetics Society of America. In 2020, Whiteman was awarded a Guggenheim Fellowship to write *Most Delicious Poison*, his first book. He was raised in northeastern Minnesota, first in Duluth and then at the edge of the vast Sax-Zim Bog wilderness, a relic of the last ice age. Trained as a botanist, an entomologist, an ornithologist, an evolutionist, and a geneticist, he began his scientific career in the Galápagos Islands, where he studied the evolution of its famous birds and their parasites. His research laboratory at Berkeley now unravels the adaptations that drive plant-animal interactions. Whiteman's pathbreaking research has been published widely in the scientific literature and featured by NPR, PBS, the *New York Times*, *Der Spiegel*, *Scientific American*, *Popular Science*, and *The Scientist*. Since 2015, Whiteman has been recognized as an Outstanding Investigator by the National Institutes of Health. He is devoted to helping create a more diverse, equitable, and just scientific research enterprise. Whiteman lives in Oakland, California, with his husband.